Advances in Magnetic and Electromagnetic Techniques for Mineral Exploration: Enhancing Resource Discovery

Advances in Magnetic and Electromagnetic Techniques for Mineral Exploration: Enhancing Resource Discovery

Guest Editors

Marc A. Vallée
Stanisław Mazur

MDPI

Basel • Beijing • Wuhan • Barcelona • Belgrade • Novi Sad • Cluj • Manchester

Guest Editors
Marc A. Vallée
Research Developments
Geo Data Solutions Inc.
Ottawa
Canada

Stanisław Mazur
Institute of Geological
Sciences
Polish Academy of Sciences
Krakow
Poland

Editorial Office
MDPI AG
Grosspeteranlage 5
4052 Basel, Switzerland

This is a reprint of the Special Issue, published open access by the journal *Minerals* (ISSN 2075-163X), freely accessible at: www.mdpi.com/journal/minerals/special_issues/XN60XS2V00.

For citation purposes, cite each article independently as indicated on the article page online and as indicated below:

Lastname, A.A.; Lastname, B.B. Article Title. *Journal Name* **Year**, *Volume Number*, Page Range.

ISBN 978-3-7258-4420-3 (Hbk)
ISBN 978-3-7258-4419-7 (PDF)
https://doi.org/10.3390/books978-3-7258-4419-7

Contents

About the Editors . . . . **vii**

Preface . . . . **ix**

Marc A. Vallée and Stanislas Azur
Advances in Magnetic and Electromagnetic Techniques for Mineral Exploration: Enhancing Resource Discovery
Reprinted from: *Minerals* **2025**, *15*, 595, https://doi.org/10.3390/min15060595 . . . . **1**

Jin Cai and Guoqing Ma
Self- Structural Constraint Joint Inversion of Aeromagnetic and Gradient Data: Enhanced Imaging for Gold Deposits in Western Henan, China
Reprinted from: *Minerals* **2025**, *15*, 337, https://doi.org/10.3390/min15040337 . . . . **4**

Junchen Liu, Jiwei Yi, Chuncheng Li and Ci Gao
Integrating Magnetic and Self-Potential Methods for Efficient Graphite Exploration: Insights from Ji'an, Northeast China
Reprinted from: *Minerals* **2025**, *15*, 27, https://doi.org/10.3390/min15010027 . . . . **23**

Ji'en Dong, Peng Wang, Hua Li, Huiqing Zhang, Mingfu Zhao, Haikui Tong, et al.
Study of the Genesis Process and Deep Prospecting Breakthrough in the Gouli Ore Concentration of the East Kunlun Metallogenic Belt Using Audio Magnetotelluric Data
Reprinted from: *Minerals* **2024**, *14*, 930, https://doi.org/10.3390/min14090930 . . . . **40**

Marc A. Vallée and Mouhamed Moussaoui
Moment Estimation from Time Domain Electromagnetic Data
Reprinted from: *Minerals* **2024**, *14*, 888, https://doi.org/10.3390/min14090888 . . . . **55**

Liu Cheng, Li Han, Yang Kai, Gao Yongbao, Tang Weidong and Yao Chuan
Location Prediction Study of Fluorite Ore in Shallow Cover Area: Evidence from Integrated Geophysical Surveys
Reprinted from: *Minerals* **2024**, *14*, 838, https://doi.org/10.3390/min14080838 . . . . **76**

Zhenyu Zhang, Yongbo Li, Liang Chen, Qiang Zhang and Yue Sun
Deep Geological Structure Analysis of the Dongyang Area, Fujian, China: Insights from Integrated Gravity and Magnetic Data
Reprinted from: *Minerals* **2024**, *14*, 837, https://doi.org/10.3390/min14080837 . . . . **96**

Alexander Prikhodko, Andrei Bagrianski and Petr Kuzmin
Airborne Natural Total Field Broadband Electromagnetics—Configurations, Capabilities, and Advantages
Reprinted from: *Minerals* **2024**, *14*, 704, https://doi.org/10.3390/min14070704 . . . . **110**

Shengping Gong, Keqiang Zhao, Mingming Wang, Shengwu Yan, Yong Li and Jianzhou Yang
Application of Integrated Geological and Geophysical Surveys on the Exploration of Chalcedony Deposits: A Case Study on Nanhong Agate in Liangshan, China
Reprinted from: *Minerals* **2024**, *14*, 677, https://doi.org/10.3390/min14070677 . . . . **126**

Marc A. Vallée, Mouhamed Moussaoui and Khorram Khan
Case Studies of Magnetic and Electromagnetic Techniques Covering the Last Fifteen Years
Reprinted from: *Minerals* **2024**, *14*, 1286, https://doi.org/10.3390/min14121286 . . . . **141**

About the Editors

Marc A. Vallée

Marc Vallée studied geological engineering and graduated from Laval University before completing a master's degree in geophysics at the University of Toronto, and a Ph.D. in geophysics at École Polytechnique de Montréal. He has over 30 years of experience as a research scientist in applied geophysics, mainly for the Noranda Technology Center, and then for Fugro Airborne Surveys and CGG companies. From 2015 to 2018, Marc was responsible for magnetic and gravity geophysical inversion as part of the NSERC-CMIC Footprints project. Since then, with AEM Expert & Marketing, he has been responsible for geophysical research and development for Geo Data Solutions GDS. His expertise is mainly in the processing and interpretation of airborne electromagnetic, magnetics, and gravity surveys. Marc has authored several scientific articles in applied geophysics.

Stanisław Mazur

Since 2018, Stanislaw Mazur has been a professor in tectonics and applied geophysics at the Institute of Geological Sciences, PAS. Mr. Mazur received his PhD in structural geology from the University of Wroclaw. Afterwards, he undertook structural geology training at the University College Dublin (1994) and spent a 2-years postdoctoral fellowship at the GeoforschungsZentrum Potsdam (20035–2004). Prior to coming to the Institute of Geological Sciences, for 11 years, he worked for British Geophysical Consulting Companies (Getech, ARKeX), where he was responsible for the technical supervision of multi-client interpretation projects. He was also a Visiting Research Fellow at the University of Leeds during 2008–2011 and 2014–2020. Mr. Mazur has authored or co-authored more than 80 peer-reviewed research papers, obtaining an h-index of 26 and c. 2000 independent citations (Web of Science). He was a member of scientific expeditions to Greenland (2001–2003), Spitsbergen (2005–2006 and 2018), British Columbia (2006), and North Labrador (2012). Since 2018, Mr. Mazur has been a head of the Research Centre in Kraków and the Depositional Systems Research Group (DEPOS).

Preface

The rapid technological progress of the 20th century significantly advanced magnetic and electromagnetic (EM) techniques, establishing them as key tools for the discovery of mineral resources. Recent advances in these methods span acquisition, sensor design, and processing. These techniques have been successfully applied in diverse geological settings worldwide, further validating their utility and expanding their scope. This special issue of *Minerals* presents a curated selection of studies that reflect current directions in magnetic and electromagnetic exploration. It particularly highlights case studies demonstrating innovative or unconventional applications of these methods in mineral discovery.

Marc A. Vallée and Stanisław Mazur
Guest Editors

minerals

Editorial

Advances in Magnetic and Electromagnetic Techniques for Mineral Exploration: Enhancing Resource Discovery

Marc A. Vallée [1,*] and Stanislaw Mazur [2]

1 Geo Data Solutions GDS Inc., Laval, QC H7P 0G1, Canada
2 Institute of Geological Sciences, Polish Academy of Sciences, 31-002 Krakow, Poland; ndmazur@cyf-kr.edu.pl
* Correspondence: mvallee@geodatasolutions.ca

1. Introduction

The rapid technological progress of the 20th century significantly advanced magnetic and electromagnetic (EM) techniques, establishing them as key tools for the discovery of mineral resources. Magnetic surveys, which detect subtle variations in the Earth's magnetic field, are particularly effective in mapping magnetite concentration [1,2]. Initially applied in the exploration of iron formations [3], magnetic techniques have evolved into powerful tools for regional geological mapping [4], and for delineating the geological context of non-magnetic mineral deposits such as gold [5].

Electromagnetic methods, by contrast, involve the use of natural or artificial electromagnetic sources and receivers that detect subsurface conductivity variations [6]. These techniques are capable of directly detecting conductive ore bodies such as massive sulfides [7], or indirectly indicating the presence of deposits like uranium through related conductivity anomalies [8].

Recent advances in these methods span several domains. In acquisition, the deployment of magnetic and EM systems on airborne platforms has been transformative [9,10]. Concurrently, improvements in sensor design, data acquisition systems, and processing techniques have enhanced sensitivity and resolution [11]. Computational developments have also led to increasingly sophisticated data interpretation workflows that integrate geophysical signatures with geological frameworks [12]. These techniques have been successfully applied in diverse geological settings worldwide, further validating their utility and expanding their scope.

This special issue presents a curated selection of studies that reflect current directions in magnetic and electromagnetic exploration. It particularly highlights case studies demonstrating innovative or unconventional applications of these methods in mineral discovery.

2. Overview of Published Articles

Contribution 1: *Cai and Ma* article introduce a novel inversion method termed *self-structural constraint (SSC)*, which enhances aeromagnetic data interpretation. Applied to a gold exploration project in western Henan, China, the SSC method aids in delineating volcanic zones with potential magmatic-hydrothermal mineralization.

Contribution 2: *Liu et al.* describe a multi-method approach combining magnetics, self-potential surveys, trenching, and drilling that led to the discovery of graphite deposits in northeastern China. Initial magnetic anomalies guided follow-up surveys and confirmed mineralization.

Contribution 3: *Dong et al.* apply the Audio-Magnetotelluric (AMT) method to the Gouli gold field, in the East Kunlun metallogenic belt, China. Two-dimensional inversions of EM data helped characterize alteration zones associated with gold mineralization, correlating well with drilling results.

Contribution 4: *Vallée and Moussaoui* propose a new method for calculating response moments from time-domain electromagnetic (TDEM) data. This approach is system agnostic and was tested using two distinct airborne TDEM systems over the Reid-Mahaffy test site, in Ontario, Canada. The technique enables comparative evaluation of system performance in identifying subsurface conductors.

Contribution 5: *Cheng et al.* report on the use of magnetic data to guide exploration for non-magnetic fluorite in the Gobi Desert, China. A low magnetic anomaly associated with structurally favorable setting was identified as a proxy for the fluorite-hosting environment.

Contribution 6: *Zhang et al.* combine gravity and magnetic to investigate deep geological structures related to gold mineralization in Fujian, China. The integrated interpretation highlighted prospective zones by correlating physical property anomalies with known geology.

Contribution 7: *Prikhodko et al.* introduce MobileMT, an innovative airborne EM system that uses natural field measurements to enhance deep penetration of the subsurface. They demonstrate its applicability across varied mineralization contexts: uranium in Canada's Athabasca Basin, massive sulfides in Ecuador, Ni-Cu deposits in Sudbury, Canada, and porphyry systems in British Columbia, Canada.

Contribution 8: *Gong et al.* employ a suite of geophysical techniques–magnetics, vertical electrical sounding, and controlled source audio-magnetotellurics– to explore agate deposits in Liangshan, China. These sedimentary host deposits lie above resistive basalt and are characterized by distinct resistivity contrasts.

Contribution 9: *Vallée et al.* present a global review of magnetic and EM applications over the last fifteen years. The authors classify studies by deposit type, location and technique, and summarize trends in tabular and graphical forms. Although biased toward North American case studies, this comprehensive synthesis serves as a practical guide for future exploration strategy development.

3. Conclusions

Magnetic and electromagnetic techniques continue to play an essential role in the evolving landscape of mineral exploration. As exploration targets diversify and move into more complex environments across all continents, technological advances–especially those integrating artificial intelligence–are expected to further revolutionize both data acquisition and interpretation. The ongoing development of these methods will be critical in meeting the global demand for mineral resources.

Conflicts of Interest: Marc A. Vallée is employee of Geo Data Solutions GDS Inc. The paper reflects the views of the scientists and not the company.

List of Contributions

1. Cai, J.; Ma, G. Self-Structural Constraint Joint Inversion of Aeromagnetic and Gradient Data: Enhanced Imaging for Gold Deposits in Western Henan, China. *Minerals* **2025**, *15*, 337.
2. Liu, J.; Yi, J.; Li, C.; Gao, C. Integrating Magnetic and Self-Potential Methods for Efficient Graphite Exploration: Insights from Ji'an, Northeast China. *Minerals* **2025**, *15*, 27.
3. Dong, J.; Wang, P.; Li, H.; Zhang, H.; Zhao, M.; Tong, H.; Yu, X.; Li, J.; Zhang, B. Study of the Genesis Process and Deep Prospecting Breakthrough in the Gouli Ore Concentration of the East Kunlun Metallogenic Belt Using Audio Magnetotelluric Data. *Minerals* **2024**, *14*, 930.

4. Vallée, M.A.; Moussaoui, M. Moment Estimation from Time Domain Electromagnetic Data. *Minerals* **2024**, *14*, 888.
5. Cheng, L.; Han, L.; Kai, Y.; Yongbao, G.; Weidong, T.; Chuan, Y. Location Prediction Study of Fluorite Ore in Shallow Cover Area: Evidence from Integrated Geophysical Surveys. *Minerals* **2024**, *14*, 838.
6. Zhang, Z.; Li, Y.; Chen, L.; Zhang, Q.; Sun, Y. Deep Geological Structure Analysis of the Dongyang Area, Fujian, China: Insights from Integrated Gravity and Magnetic Data. *Minerals* **2024**, *14*, 837.
7. Prikhodko, A.; Bagrianski, A.; Kuzmin, P. Airborne Natural Total Field Broadband Electromagnetics—Configurations, Capabilities, and Advantages. *Minerals* **2024**, *14*, 704.
8. Gong, S.; Zhao, K.; Wang, M.; Yan, S.; Li, Y.; Yang, J. Application of Integrated Geological and Geophysical Surveys on the Exploration of Chalcedony Deposits: A Case Study on Nanhong Agate in Liangshan, China. *Minerals* **2024**, *14*, 677.
9. Vallée, M.A.; Moussaoui, M.; Khan, K. Case Studies of Magnetic and Electromagnetic Techniques Covering the Last Fifteen Years. *Minerals* **2024**, *14*, 1286.

References

1. Nabighian, M.N.; Grauch, V.J.S.; Hansen, R.O.; LaFehr, T.R.; Li, Y.; Peirce, J.W.; Phillips, J.D.; Ruder, M.E. The historical development of the magnetic method in exploration. *Geophysics* **2005**, *70*, 33ND–61ND. [CrossRef]
2. Grant, F.S. Aeromagnetics, geology and ore environments, I. Magnetite in igneous, sedimentary and metamorphic rocks: An overview. *Geoexploration* **1985**, *23*, 303–333. [CrossRef]
3. Jahren, C.E. Magnetic susceptibility of bedded iron-formation. *Geophysics* **1963**, *28*, 756–766. [CrossRef]
4. Hinze, W.J. (Ed.) *The Utility of Regional Gravity and Magnetic Anomaly Maps*; Society of Exploration Geophysicists: Tulsa, OK, USA, 1985.
5. Doyle, H.A. Geophysical exploration for gold—A review. *Explor. Geophys.* **1986**, *17*, 169–180. [CrossRef]
6. Grant, F.S.; West, G.F. *Interpretation Theory in Applied Geophysics*; McGraw-Hill Book Company: New York, NY, USA, 1965.
7. Morgan, L.A. Geophysical Characteristics of Volcanogenic Massive Sulfide Deposits. In *Volcanogenic Sulfide Occurrence Model*; Shanks, P., Thurston, R., Eds.; US Geological Survey: Reston, VA, USA, 2012; pp. 113–131.
8. Tuncer, V.; Unsworth, M.J.; Siripunvaraporn, W.; Craven, J.A. Exploration for unconformity-type uranium deposits with audiomagnetotelluric data: A case study from the McArthur River mine, Saskatchewan, Canada. *Geophysics* **2006**, *71*, B201–B209. [CrossRef]
9. Hood, P. History of aeromagnetic surveying in Canada. *Lead. Edge* **2007**, *26*, 1384–1392. [CrossRef]
10. Fountain, D. Airborne electromagnetic systems—50 years of development. *Explor. Geophys.* **1998**, *29*, 1–11. [CrossRef]
11. Macnae, J. Advances in Electromagnetic Data Processing: Noise, Signal, SPM and AIP. In Proceedings of the Exploration 17: Sixth Decennial International Conference on Mineral Exploration, Toronto, ON, Canada, 22–25 October 2017; Tschirhart, V., Thomas, M.D., Eds.; Decennial Mineral Exploration Conferences: Toronto, ON, Canada, 2017; pp. 191–208.
12. Farquharson, C.G.; Lelièvre, P.G. Modelling and Inversion for Mineral Exploration Geophysics: A Review of Recent Progress, the Current State-of-the-Art, and Future Directions. In Proceedings of the Exploration 17: Sixth Decennial International Conference on Mineral Exploration, Toronto, ON, Canada, 22–25 October 2017; Tschirhart, V., Thomas, M.D., Eds.; Decennial Mineral Exploration Conferences: Toronto, ON, Canada, 2017; pp. 51–74.

minerals

MDPI

Article

Self-Structural Constraint Joint Inversion of Aeromagnetic and Gradient Data: Enhanced Imaging for Gold Deposits in Western Henan, China

Jin Cai [1,2,*] and Guoqing Ma [2]

[1] College of Mathematics and Physics Science, Hunan University of Arts and Science, Changde 415000, China
[2] College of Geoexploration Science and Technology, Jilin University, Changchun 130026, China; maguoqing@jlu.edu.cn
* Correspondence: caijin@huas.edu.cn

Abstract: Innovative magnetic techniques are pivotal for advancing mineral exploration. This study presents a self-structural constraint (SSC) method that jointly inverts aeromagnetic and gradient data to resolve high-resolution magnetic susceptibility models for concealed ores. The SSC framework integrates gradient structures from multi-component data as mutual constraints, enhancing signal differentiation and noise suppression. Unstructured tetrahedral grids and Poisson-derived analytical expressions address complex terrains, enabling robust inversions. Synthetic tests show SSC improves resolution by 40%–60% over conventional methods and resists 10% Gaussian noise. Applied to gold exploration in western Henan, China, SSC delineated concealed ore bodies (300–2000 m depth) along NE- and NW-trending faults, correlating with andesite-hosted magnetic anomalies. Combined with volcanic facies analysis, magma migration through these faults provided metallogenic materials and structural traps. The SSC-derived 3D model identified new drill targets, bridging geophysical imaging with geological processes. This advancement enhances the detection of deep, structurally controlled mineralization, offering a transformative tool for resource discovery.

Keywords: mineral exploration; 3D magnetic imaging; self-structural constraint (SSC); aeromagnetic gradient inversion; magnetic susceptibility modeling; concealed ore detection

1. Introduction

Three-dimensional inversion of magnetic data has emerged as an indispensable methodology for characterizing subsurface magnetic susceptibility distributions, offering critical insights into the exploration of magnetic mineral resources [1–7]. Building upon the foundational framework established by Li and Oldenburg [8] for 3D magnetic inversion, subsequent advancements in computational efficiency, memory optimization, and resolution enhancement [9–11] have significantly broadened the scope of applications. These include, but are not limited to, mineral resource delineation [5,7], crustal architecture reconstruction [12], and detection of unexploded ordnance [13].

Magnetic gradient data exhibit inherently superior horizontal resolution [14], enabling enhanced delineation of near-surface magnetic sources. Building on this capability, recent advancements in joint inversion frameworks—integrating total-field and gradient magnetic data [14–16]—have significantly improved subsurface imaging accuracy by systematically leveraging multi-component datasets through unified computational matrices.

The inherent complexity of natural terrains—characterized by undulating topography and irregular subsurface geological bodies—challenges the efficacy of traditional inversion methods. To overcome these limitations, unstructured discretization techniques grounded in constrained Delaunay triangulation have emerged as robust alternatives [17]. In a seminal study, Abedi [18] developed a 2D focusing inversion method tailored for potential field data in rugged terrains, demonstrating enhanced stability for abrupt geological transitions. Meanwhile, Zuo et al. [19] advanced the field by proposing a 3D accelerated magnetic inversion algorithm that synergizes partial differential equations with unstructured tetrahedral meshing, effectively addressing computational bottlenecks in geometrically complex scenarios.

Recent advances in self-constrained inversion methodologies have demonstrated significant progress in addressing geophysical interpretation challenges. Paoletti et al. [20] pioneered a potential-field-constrained inversion framework utilizing a priori information derived exclusively from gravity and magnetic data analysis, establishing the foundational concept of self-constrained inversion. Building on this concept, Davide et al. [21] integrated microgravity surveys with a self-constrained inversion strategy to resolve the shallow geometry of the Irpinia Fault in Southern Italy, concurrently estimating its Holocene slip rate validated against independent geological constraints. Similarly, Sun and Chen [22] developed a magnetic gradient-driven self-constrained approach, where cross-correlation analyses between theoretical and observed data gradients were translated into spatial weighting functions to reinforce model geological plausibility. In parallel, Vitale and Fedi [23] introduced a two-step self-consistent inversion paradigm for potential fields, synergizing multiscale homogeneity analysis with 3D variable-exponent depth-weighting derived from field scaling properties, thereby enhancing the reconstruction fidelity of deep-seated complex sources. The integration of machine learning has further expanded methodological boundaries: Zhou et al. [24] devised a dual-network self-constrained architecture with adaptive fine-tuning to augment gravity inversion precision, effectively bridging data-driven learning with physical constraints, as evidenced by its successful application in geothermal reservoir delineation. Complementarily, Ming et al. [25] formulated a power-type structural self-constrained inversion (PTSS) scheme incorporating L2-norm regularization, where power-gradient self-constraints and cross-physics mutual constraints synergistically sharpened boundary resolution in joint gravity-magnetic inversions, a capability decisively validated in iron ore exploration scenarios.

In this study, we introduce a self-structural constraint (SSC) inversion method designed to significantly enhance the resolution of joint total-field and gradient magnetic data inversion. The SSC framework uniquely integrates gradient-derived structural attributes—extracted from multi-component inversion results—as physical constraints, thereby amplifying the differentiation of subsurface signals and suppressing noise interference. To address the challenges posed by complex terrains, we derive analytical expressions for tetrahedral magnetic gradient anomalies rooted in Poisson's theory, enabling robust SSC inversion within an adaptive unstructured tetrahedral meshing framework. Theoretical model validations and real-world case studies demonstrate that the SSC method achieves 40%–60% higher spatial resolution in magnetic susceptibility imaging compared to conventional approaches while maintaining stability under noise levels up to 20% SNR. This advancement establishes SSC as a transformative tool for high-fidelity subsurface exploration in geologically intricate environments.

2. Methodology

Unstructured tetrahedral discretization provides an optimal framework for modeling undulating terrains and irregular geological geometries, enabling the derivation of tetrahedral magnetic gradient forward formulas based on Poisson's theory to facilitate SSC inversion within an adaptive meshing framework.

2.1. Forward of Gradient Magnetic Anomaly of Tetrahedron

The expression of total-field magnetic anomaly ($\boldsymbol{d}_{\Delta T}$) is

$$\begin{aligned} \boldsymbol{d}_{\Delta T} &= \mathbf{G}_{\Delta T} \cdot \boldsymbol{\kappa} \\ &= \tfrac{T_0}{4\pi G\rho}\{(V_{xx}\cos i\cos d + V_{xy}\cos i\sin d + V_{xz}\sin i)\cdot\cos I\cos D + \\ &(V_{xy}\cos i\cos d + V_{yy}\cos i\sin d + V_{yz}\sin i)\cdot\cos I\sin D + \\ &(V_{xz}\cos i\cos d + V_{yz}\cos i\sin d + V_{zz}\sin i)\cdot\sin I\}\cdot\boldsymbol{\kappa} \end{aligned} \quad (1)$$

where $\boldsymbol{G}_{\Delta T}$ is the kernel matrix connecting the vector of total-field magnetic anomaly data ($\boldsymbol{d}_{\Delta T}$) and the vector of magnetic susceptibility ($\boldsymbol{\kappa}$), which is calculated using a forward analytical expression. T_0 is the geomagnetic field intensity, G is the gravitational constant, and ρ is the density of the geological body. I and D are the inclinations and declinations of the geomagnetic field, and i and d are the inclinations and declinations of the geological body, respectively. x, y, and z axes indicate easting, northing, and vertically downward, respectively. Analytical expressions for the second-order derivatives of the gravity potential (V_{xx}, V_{xy}, V_{xz}, V_{yy}, V_{yz}, and V_{zz}) of the tetrahedron are provided in Appendix A.

We derived the forward analytical expressions of magnetic gradient anomalies with unstructured tetrahedral grid meshing based on Poisson's theory. The magnetic gradient anomalies in three directions ($\boldsymbol{d}_{\Delta T_x}$, $\boldsymbol{d}_{\Delta T_y}$, and $\boldsymbol{d}_{\Delta T_z}$) can be expressed as

$$\begin{aligned} \boldsymbol{d}_{\Delta T_x} &= \mathbf{G}_{\Delta T_x} \cdot \boldsymbol{\kappa} \\ &= \tfrac{T_0}{4\pi G\rho}\{(V_{xxx}\cos i\cos d + V_{xxy}\cos i\sin d + V_{xxz}\sin i)\cdot\cos I\cos D + \\ &(V_{yxx}\cos i\cos d + V_{yxy}\cos i\sin d + V_{yxz}\sin i)\cdot\cos I\sin D + \\ &(V_{zxx}\cos i\cos d + V_{zxy}\cos i\sin d + V_{zxz}\sin i)\cdot\sin I\}\cdot\boldsymbol{\kappa} \end{aligned} \quad (2)$$

$$\begin{aligned} \boldsymbol{d}_{\Delta T_y} &= \mathbf{G}_{\Delta T_y} \cdot \boldsymbol{\kappa} \\ &= \tfrac{T_0}{4\pi G\rho}\{(V_{yxx}\cos i\cos d + V_{yxy}\cos i\sin d + V_{yxz}\sin i)\cdot\cos I\cos D + \\ &(V_{yxy}\cos i\cos d + V_{yyy}\cos i\sin d + V_{yyz}\sin i)\cdot\cos I\sin D + \\ &(V_{zxy}\cos i\cos d + V_{zyy}\cos i\sin d + V_{zyz}\sin i)\cdot\sin I\}\cdot\boldsymbol{\kappa} \end{aligned} \quad (3)$$

$$\begin{aligned} \boldsymbol{d}_{\Delta T_z} &= \mathbf{G}_{\Delta T_z} \cdot \boldsymbol{\kappa} \\ &= \{(V_{zxx}\cos i\cos d + V_{zxy}\cos i\sin d + V_{zxz}\sin i)\cdot\cos I\cos D + \\ &(V_{zxy}\cos i\cos d + V_{zyy}\cos i\sin d + V_{zyz}\sin i)\cdot\cos I sin D + \\ &(V_{zxz}\cos i\cos d + V_{zyz}\cos i\sin d + V_{zzz}\sin i)\cdot\sin I\}\cdot\boldsymbol{\kappa} \end{aligned} \quad (4)$$

where $\boldsymbol{G}_{\Delta T_x}$, $\boldsymbol{G}_{\Delta T_y}$, and $\boldsymbol{G}_{\Delta T_z}$ are the kernel matrices connecting the vectors of 3D magnetic gradient anomaly data ($\boldsymbol{d}_{\Delta T_x}$, $\boldsymbol{d}_{\Delta T_y}$ and $\boldsymbol{d}_{\Delta T_z}$) and the vector of magnetic susceptibility ($\boldsymbol{\kappa}$), which were calculated by forward analytical expressions. We used the coordinate transformation provided by Okabe [26] to derive the analytical expressions of the third-order derivative of the gravity potential (V_{xyz}, V_{zzy}, V_{zzz}, V_{zzx}, V_{xxy}, V_{xxz}, V_{yyx}, V_{yyz}, V_{xxx} and V_{yyy}) of the tetrahedron, which are provided in Appendix B.

To validate the accuracy of the derived forward analytical expressions for tetrahedral total-field and gradient magnetic anomalies, we constructed a regular hexahedral model with dimensions of 300 × 300 × 300 m (Figure 1a) and discretized it into 12 tetrahedral elements of varying sizes using Delaunay triangulation (Figure 1b_1–1b_{12}).

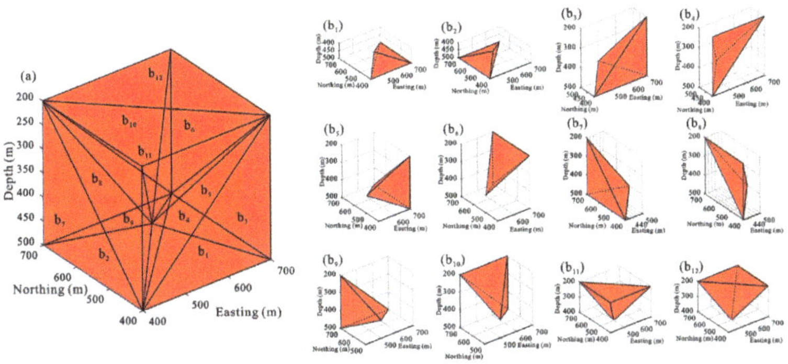

Figure 1. (**a**) Schematic of the regular hexahedron validation model (b_1–b_{12} are 12 tetrahedrons of different sizes obtained after Delaunay triangulation). (**b**) Tetrahedral meshing of the regular hexahedron partitioned into 12 unstructured tetrahedrons via Delaunay triangulation.

The forward analytical expressions were employed to compute the total-field and gradient magnetic anomalies for the 12 tetrahedral elements derived from the subdivision of the regular hexahedron (Figure 2a–d). These results were directly compared with those obtained using conventional hexahedral forward formulas (Figure 2e–h), with the geomagnetic field parameters set to an inclination of 45° and a declination of 60°. The subtraction of Figure 2a–d from Figure 2e–h, respectively, yields the anomalous residual diagrams shown in Figure 2i–l, where all residuals are close to zero. Subsequently, the root mean square errors (RMSEs) between the total-field and gradient magnetic anomalies calculated by the tetrahedral forward analytical expression and those directly computed using the hexahedral forward formula were evaluated. The RMSE between Figure 2a,e is 5.8395×10^{-4}, between Figure 2b,f is 4.5219×10^{-6}, between Figure 2c,g is 3.4496×10^{-6}, and between Figure 2d,h is 6.7600×10^{-6}. These results further validate the accuracy of the tetrahedral forward analytical expression, thereby laying the foundation for subsequent joint inversion of the magnetic total-field and its gradient under unstructured tetrahedral mesh discretization.

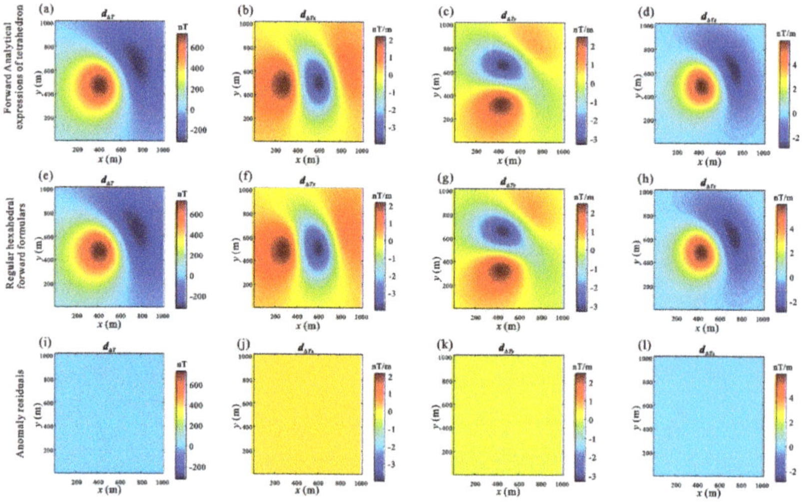

Figure 2. **(a–d)** $\boldsymbol{d}_{\Delta T}$, $\boldsymbol{d}_{\Delta T_x}$, $\boldsymbol{d}_{\Delta T_y}$, and $\boldsymbol{d}_{\Delta T_z}$ calculated by the sum of 12 tetrahedrons. **(e–h)** $\boldsymbol{d}_{\Delta T}$, $\boldsymbol{d}_{\Delta T_x}$, $\boldsymbol{d}_{\Delta T_y}$, and $\boldsymbol{d}_{\Delta T_z}$ calculated by the analytical expressions of a hexahedron. **(i–l)** Subtract **(a–d)** from **(e–h)**, respectively.

2.2. SSC Method

The conventional joint inversion of total-field and gradient magnetic data involves placing all data in a matrix, and the objective function Φ_{Ji} is

$$\begin{aligned}\Phi_{Ji} &= \Phi_d(\boldsymbol{\kappa}) + \delta\Phi_m(\boldsymbol{\kappa}) \\ &= \left\| \left(\begin{bmatrix} \boldsymbol{G}_{\Delta T} \\ \boldsymbol{G}_{\Delta T_x} \\ \boldsymbol{G}_{\Delta T_y} \\ \boldsymbol{G}_{\Delta T_z} \end{bmatrix} \boldsymbol{\kappa} - \begin{bmatrix} \boldsymbol{d}_{\Delta T} \\ \boldsymbol{d}_{\Delta T_x} \\ \boldsymbol{d}_{\Delta T_y} \\ \boldsymbol{d}_{\Delta T_z} \end{bmatrix} \right) \right\|_2^2 + \delta \left\| \boldsymbol{W}_{m_{Ji}} \boldsymbol{\kappa} \right\|_2^2 \to \min\end{aligned} \tag{5}$$

where $\boldsymbol{W}_{m_{Ji}} = diag\left(\left(\begin{bmatrix} \boldsymbol{G}_{\Delta T} \\ \boldsymbol{G}_{\Delta T_x} \\ \boldsymbol{G}_{\Delta T_y} \\ \boldsymbol{G}_{\Delta T_z} \end{bmatrix} \right)^T \cdot \begin{bmatrix} \boldsymbol{G}_{\Delta T} \\ \boldsymbol{G}_{\Delta T_x} \\ \boldsymbol{G}_{\Delta T_y} \\ \boldsymbol{G}_{\Delta T_z} \end{bmatrix} \right)^{\frac{1}{2}}$ is the model weighting matrix, and δ is the regularization parameter [8,15,27].

To enhance the resolution of the joint inversion of total-field and gradient magnetic data (Equation (5)), we propose a self-structural constraint (SSC) method. This approach leverages gradient-derived structural features from the inversion results of both total-field and gradient magnetic data as physical constraints. By integrating these constraints, the SSC method effectively combines the deep-source recovery capability of total-field magnetic data inversion with the high-resolution imaging advantages of magnetic gradient data inversion.

The objective function of the SSC method is

$$\Phi_{SSC} = \left\| \left(\begin{bmatrix} \boldsymbol{G}_{\Delta T} \\ \boldsymbol{G}_{\Delta T_x} \\ \boldsymbol{G}_{\Delta T_y} \\ \boldsymbol{G}_{\Delta T_z} \end{bmatrix} \boldsymbol{\kappa}_3 - \begin{bmatrix} \boldsymbol{d}_{\Delta T} \\ \boldsymbol{d}_{\Delta T_x} \\ \boldsymbol{d}_{\Delta T_y} \\ \boldsymbol{d}_{\Delta T_z} \end{bmatrix} \right) \right\|_2^2 + \delta \| \boldsymbol{W}_{m_{SSC}} \boldsymbol{\kappa}_3 \|_2^2 + \gamma_{SSC} \left(\| \boldsymbol{t}_1 \|_2^2 + \| \boldsymbol{t}_2 \|_2^2 \right) \to \min \tag{6}$$

The solution process differentiated Equation (6) with respect to κ_{3W} and can be expressed as

$$\frac{\partial \Phi_{SSC}}{\partial \kappa_{3w}} = \Big\{ G_{3w}{}^T G_{3w} + \alpha E + \gamma_{SSC} \Big[\left(B_{xw}^{\kappa_1}\right)^T B_{xw}^{\kappa_1} + \left(B_{yw}^{\kappa_1}\right)^T B_{yw}^{\kappa_1} + \left(B_{zw}^{\kappa_1}\right)^T B_{yw}^{\kappa_1} + \left(B_{xw}^{\kappa_2}\right)^T B_{xw}^{\kappa_2} + \left(B_{yw}^{\kappa_2}\right)^T B_{yw}^{\kappa_2} + \left(B_{zw}^{\kappa_2}\right)^T B_{yw}^{\kappa_2} \Big] \Big\} \kappa_{3w} - G_{3w}^T d_3 \tag{7}$$

where $G_{3w} = \begin{bmatrix} G_{\Delta T} \\ G_{\Delta T_x} \\ G_{\Delta T_y} \\ G_{\Delta T_z} \end{bmatrix} \cdot W_{m_{SSC}}, d_3 = \begin{bmatrix} d_{\Delta T} \\ d_{\Delta T_x} \\ d_{\Delta T_y} \\ d_{\Delta T_z} \end{bmatrix}$. E is an identity matrix and the optimal solution of the objective function transformed into Equation (7) is zero. γ_{SSC} is a regularization parameter. κ_{3w} is the weighting physical property, $\kappa_{3w} = \kappa_3 \cdot W_{m_{SSC}}$, $W_{m_{SSC}}$ is the weight of magnetic susceptibility,

$$W_{m_{SSC}} = \left(\sum_{i=1}^{Q} \left(\begin{bmatrix} G_{\Delta T} \\ G_{\Delta T_x} \\ G_{\Delta T_y} \\ G_{\Delta T_z} \end{bmatrix} \right)^2_{ij} \right)^{\frac{\varepsilon}{4}} \cdot diag\left[{(S_W - \min(S_W))} \Big/ {(\max(S_W) - \min(S_W))} \right], j = 1, \ldots, P,$$

Q is the number of observation points, and P is the number of tetrahedral units. $S_W = {(\kappa_1 + \kappa_2)} \big/ 2$, κ_1, and κ_2 are the inversion results of the total-field and gradient magnetic data, respectively. $\varepsilon(0.5 \le \varepsilon \le 1.5)$ is the depth weighting factor, which is usually taken as 1. diag() converts the vector in parentheses into a diagonal matrix.

In Equation (6), t_1 and t_2 are the gradient constraint terms of the inversion results for the total-field and gradient magnetic anomalies, respectively. The calculation formulas are as follows:

$$\|t_1\|_2^2 = \left(B_{xw}^{\kappa_1} \kappa_{3W}\right)^T \left(B_{xW}^{\kappa_1} \kappa_{3W}\right) + \left(B_{yW}^{\kappa_1} \kappa_{3W}\right)^T \left(B_{yW}^{\kappa_1} \kappa_{3W}\right) + \left(B_{zW}^{k_1} k_{3W}\right)^T \left(B_{zW}^{k_1} k_{3W}\right) \tag{8}$$

$$\|t_2\|_2^2 = \left(B_{xW}^{k_2} k_{3W}\right)^T \left(B_{xW}^{k_2} k_{3W}\right) + \left(B_{yW}^{k_2} k_{3W}\right)^T \left(B_{yW}^{k_2} k_{3W}\right) + \left(B_{zW}^{k_2} k_{3W}\right)^T \left(B_{zW}^{k_2} k_{3W}\right) \tag{9}$$

where $B_{xw}^{k_1} = B_x^{k_1} \cdot W_{m_1}^{-1}$, $B_{xw}^{k_2} = B_x^{k_2} \cdot W_{m_2}^{-1}$, $B_{yw}^{k_1} = B_y^{k_1} \cdot W_{m_1}^{-1}$, $B_{yw}^{k_2} = B_y^{k_2} \cdot W_{m_2}^{-1}$, $B_{zw}^{k_1} = B_z^{k_1} \cdot W_{m_1}^{-1}$, $B_{zW}^{k_2} = B_z^{k_2} \cdot W_{m_2}^{-1}$, $W_{m_i} = diag\left(\left(w_h^{\varepsilon}\right)^{-1} \cdot w_V^{\lambda} \right)$, $i = 1, 2$, w_h and w_V are the depth and volume of each tetrahedron unit, respectively. λ is a volume weight factor, which is generally 0.5. $B_x^{k_i}$, $B_y^{k_i}$, and $B_z^{k_i}$ are the gradient matrices of the physical properties (κ_1 and κ_2) in the x, y, and z directions, respectively, and the gradient calculation with unstructured tetrahedral grid meshing was obtained using the second-order Taylor formula [28].

3. Theoretical Model Tests

To simulate realistic subsurface conditions, we designed two inclined prism models with varying burial depths, set within a geomagnetic field characterized by an inclination of 45° and a declination of 60°. The detailed model parameters are provided in Table 1.

Table 1. Model parameters.

Model	Center Coordinates of the Top (m)	Center Coordinates of the Bottom (m)	Length of Top and Bottom (m)	Width of Top and Bottom (m)	Magnetic Susceptibility (SI)
Inclined prism 1	(1700, 1300, −300)	(2000, 1400, −1300)	600	500	0.025
Inclined prism 2	(2000, 2900, −500)	(1700, 2800, −1600)	850	750	0.025

Figure 3a shows the spatial distribution of the models and the total-field magnetic anomaly ($d_{\Delta T}$). Figure 3b–d show the magnetic gradient anomalies in three directions. Inversion calculations of the total-field and gradient magnetic anomalies were carried out, and the sections of the 3-D inversion results with an easting of 2000 m were obtained. The white solid boxes are the true positions of the models with an easting of 2000 m (Figure 3e–h). The inversion result obtained by the 3D magnetic gradient anomalies (Figure 3f) better depicted the upper boundary of the geological body and the shallower geological body. Thus, we conducted joint inversion of the total-field magnetic anomaly and 3D magnetic gradient anomalies (Figure 3g) and compared it with the inversion result of the total-field magnetic anomaly (Figure 3e). It was inferred that the joint inversion strategy would provide more information on deeper geological bodies. The inversion result obtained by the SSC method (Figure 3h) had a higher resolution and clearer boundaries, indicating that the introduction of gradient structures of physical properties can significantly improve the recovery ability of the magnetic susceptibility distribution of subsurface magnetic bodies. Under identical conditions, the recovered magnetic susceptibility values are 2.6 times higher than those from the joint inversion of total-field magnetic data and its gradients, with a 1.6-fold improvement in resolution. Since the iteration-stopping criteria were met at the 100th iteration, the entire inversion process iterated 100 times, and the root mean square (RMS) fitting error was calculated for each iteration. The final RMS fitting error for the joint inversion of total-field magnetic data and its gradients was 0.15 nT, whereas the self-structural constrained joint inversion of total-field magnetic data and its gradients achieved a final RMS fitting error of 0.02 nT. These results further demonstrate that the proposed self-structural constrained joint inversion method significantly enhances the accuracy of subsurface source recovery, as evidenced by the markedly reduced fitting error.

To simulate realistic field conditions where measured total-field and gradient magnetic anomalies are typically contaminated by noise, we introduced Gaussian noise with a signal-to-noise ratio (SNR) of 10 to the anomalies shown in Figure 3a–d. The resulting noisy total-field magnetic anomaly and 3D magnetic gradient anomalies are presented in Figure 4a–d, respectively. Figure 4e,f display cross-sections of the 3D inversion results at an Easting of 2000 m, obtained using conventional joint inversion and the SSC method, with white solid boxes indicating the true model positions at this easting. Both methods demonstrated robust noise immunity, effectively recovering the true model distribution with accuracy comparable to noise-free inversion scenarios. In addition to the reported 10% Gaussian noise test, the SSC method was evaluated under varying noise levels (5%, 15%, and 20% SNR). At 5% noise, SSC maintained >90% recovery accuracy for shallow and mid-depth structures. At 15%–20% noise, resolution improvement remained significant (30%–50% over conventional methods), though minor artifacts emerged in deeper regions (>1500 m). The proposed method in this study recovers magnetic susceptibility values four times higher than those obtained from the joint inversion of total-field magnetic data and its gradients under identical conditions, with a threefold improvement in resolution. For comparative purposes, the inversion process was set to 100 iterations, and the root mean square (RMS) fitting error was calculated at each iteration. The final RMS fitting error for the joint inversion of total-field magnetic data and its gradients was 13.33 nT, while the self-structural constrained joint inversion achieved a final RMS fitting error of 0.34 nT. Due to the influence of noise on the anomalous data, the inversion results partially fit the noise, leading to an increase in the RMS fitting error compared to noise-free conditions. This further highlights the robustness and superior accuracy of the self-structural constrained joint inversion method in recovering subsurface sources under noisy scenarios.

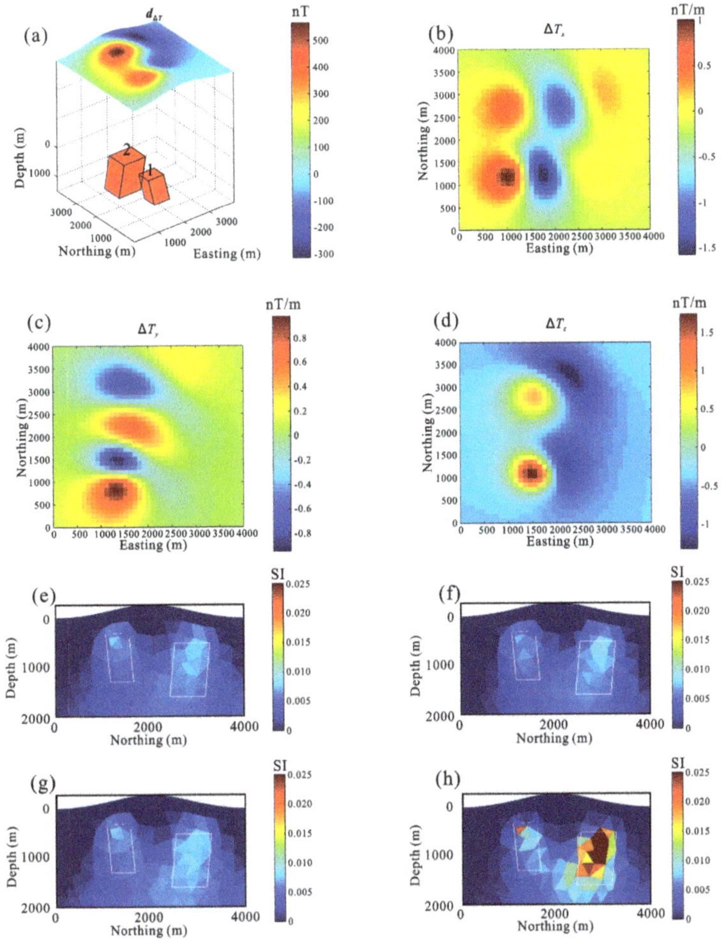

Figure 3. (**a**) Synthetic inclined prism models and their total-field magnetic anomaly. Magnetic gradient anomaly in (**b**) x-direction, (**c**) y-direction, and (**d**) z-direction. Section of the 3D inversion result of (**e**) total-field magnetic anomaly with easting of 2000 m and (**f**) 3D magnetic gradient anomalies with easting of 2000 m. (**g**) Section of the 3D conventional joint inversion result of total-field magnetic anomaly and 3D magnetic gradient anomalies with easting of 2000 m. (**h**) Section of the 3D joint inversion result of total-field magnetic anomaly and 3D magnetic gradient anomalies by the SSC method with easting of 2000 m.

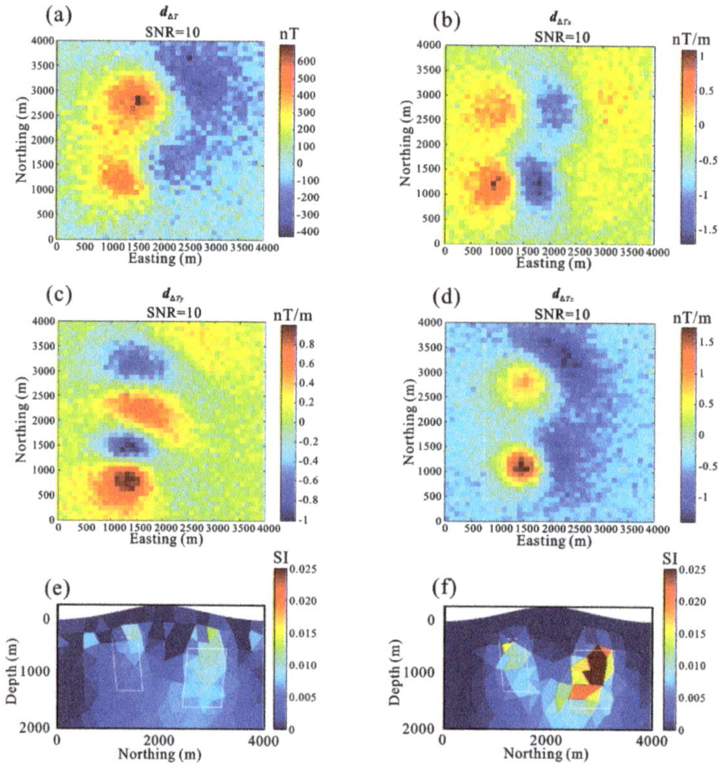

Figure 4. (**a**) Total-field magnetic anomaly (Gaussian noise with SNR of 10). Magnetic gradient anomaly (Gaussian noise with SNR of 10) in (**b**) x-direction, (**c**) y-direction, and (**d**) z-direction. (**e**) Section of the 3D conventional joint inversion result of total-field magnetic anomaly and 3D magnetic gradient anomalies with easting of 2000 m. (**f**) Section of the 3D joint inversion result of total-field magnetic anomaly and 3D magnetic gradient anomalies by the SSC method with easting of 2000 m.

In realistic geological scenarios, subsurface magnetic bodies often exhibit complex distributions. To evaluate the applicability of the SSC method under such conditions, we designed three inclined prism models with identical burial depths (Figure 5), whose specific parameters are detailed in Table 2.

Table 2. Model parameters.

Model	Center Coordinates of the Top (m)	Center Coordinates of the Bottom (m)	Length of Top and Bottom (m)	Width of Top and Bottom (m)	Magnetic Susceptibility (SI)
Inclined prism 1	(1300, 1300, −300)	(1400, 1400, −1300)	1000	1000	0.25
Inclined prism 2	(2900, 1300, −300)	(2800, 1400, −1300)	1000	1000	0.25
Inclined prism 3	(2900, 2900, −300)	(2800, 2800, −1300)	1000	1000	0.25

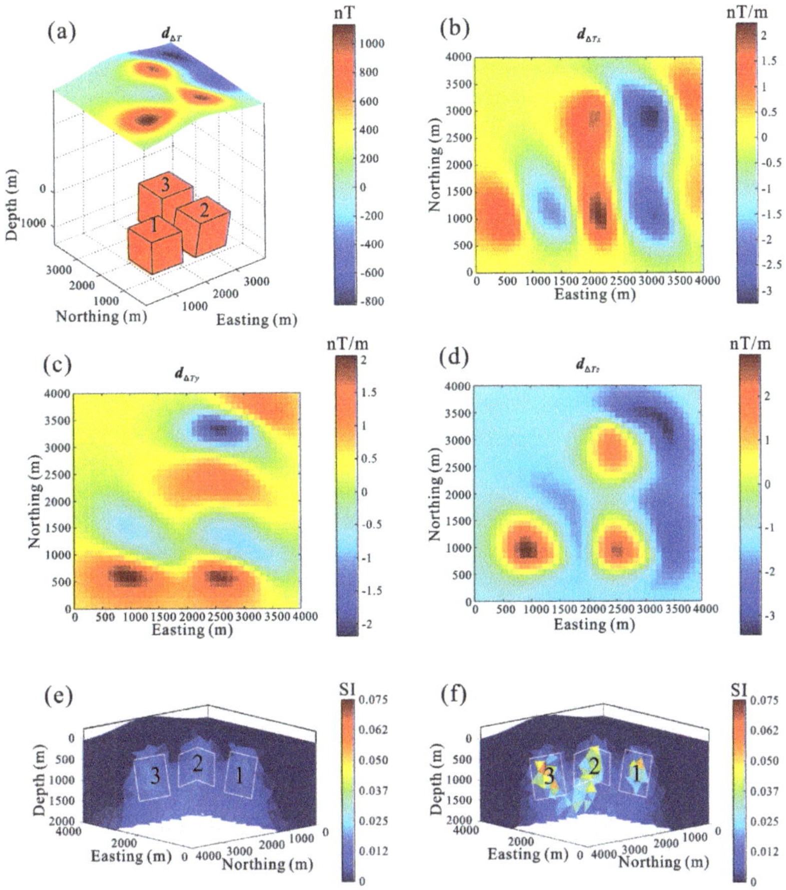

Figure 5. (**a**) Inclined prism models and total-field magnetic anomaly. Magnetic gradient anomaly in (**b**) x-direction, (**c**) y-direction, and (**d**) z-direction. (**e**) Conventional joint inversion result of total-field magnetic anomaly and 3D magnetic gradient anomalies in the area where easting > 2800 m or northing > 2800 m (The numbers 1–3 correspond to the inclined prisms 1–3 in figure (**a**)). (**f**) Three-dimensional joint inversion result of total-field magnetic anomaly and 3D magnetic gradient anomalies by the SSC method in area where easting > 2800 m or northing > 2800 m.

Figure 5a illustrates the spatial distribution of the models and their corresponding total-field magnetic anomaly, while Figure 5b–d present the magnetic gradient anomalies along the three orthogonal directions ($\boldsymbol{d}_{\Delta T_x}$, $\boldsymbol{d}_{\Delta T_y}$, and $\boldsymbol{d}_{\Delta T_z}$). The 3-D inversion results for the region where easting > 2800 m or northing > 2800 m are shown in Figure 5e,f, with white solid boxes indicating the true positions of the models within this area. Compared to the conventional joint inversion results (Figure 5e), the SSC-based joint inversion (Figure 5f) demonstrated significantly improved resolution, with recovered magnetic susceptibility distributions more closely approximating the true model geometries. The proposed method in this study recovers magnetic susceptibility values 3.8 times higher than those from the joint inversion of total-field magnetic data and its gradients under identical conditions, with a 2.8-fold improvement in resolution. During the inversion process, the iteration count was set to 100, and the root mean square (RMS) fitting error was calculated at each iteration. The final RMS fitting error for the conventional joint inversion of total-field magnetic data and its gradients was 6.09 nT, whereas the self-structural constrained joint inversion achieved a final RMS fitting error of 0.76 nT. These results demonstrate that, for more complex subsurface scenarios, the proposed self-structural constrained joint inversion method yields significantly higher accuracy in inversion results and better recovers the magnetic structures of subsurface magnetic bodies.

4. Real Data Application

The study area is situated in the southern sector of the ore district within the western mountainous region of Songxian County, Henan Province, China. As illustrated on the right side of Figure 6, the geological and tectonic framework of the area is characterized by a predominance of fault structures, with limited fold development and NE-dipping monoclinic strata [29]. Multiphase tectonic activity involving stresses from varying directions has generated a complex fracture network throughout the region. This structural complexity is further manifested in the formation of mylonite belts, cataclastic rock zones, and altered cataclastic rock belts, which result from ductile–brittle deformation and metamorphic processes. These features exhibit a strong spatial correlation with gold mineralization. The gold deposits, labeled as I, II, III, and IV in Figure 6, are strategically positioned within this structurally controlled metallogenic system.

The left side of Figure 6 presents a detailed profile (corresponding to the solid white line on the right side of Figure 6) of gold deposit I within the study area. The exposed strata belong to the upper section of the Middle Proterozoic Great Wall System, predominantly comprising volcanic and tectonic breccia. Among the volcanic rocks, andesite is the most extensively distributed and exhibits strong magnetic properties. In contrast, tectonic breccia, being a sedimentary rock, is generally non-magnetic or weakly magnetic. The blue dotted circle indicates the approximate location and spatial extent of the gold orebody. Drill hole data, spanning depths of 21.22–759.57 m, confirm the presence of gold mineralization. The ore-bearing structures exhibit undulating geometries along their dip, with dip angles progressively varying from surface to depth. Notably, gold mineralization is predominantly concentrated in transitional zones where the dip angle shifts from steep to shallow, highlighting the structural control of ore localization.

Figure 7 displays the topographic map of the study area, which is characterized by a highly mountainous terrain. The region exhibits a distinct geomorphological pattern, with elevated northern and southern sectors and a relatively lower central zone, creating significant topographic variations. From south to north, the area is dominated by three primary geomorphological features: Xiong'er Mountain, Yihe River Valley, and Waifang Mountain. The Yihe River Valley serves as a natural boundary separating Xiong'er Mountain to the south from Waifang Mountain to the north. Topographic elevations within the study area range from 300 to 800 m, reflecting the rugged nature of the landscape. The presence of trenches in the western portion of the study area is manifested as linear features on the topographic map, adding to the complexity of the terrain.

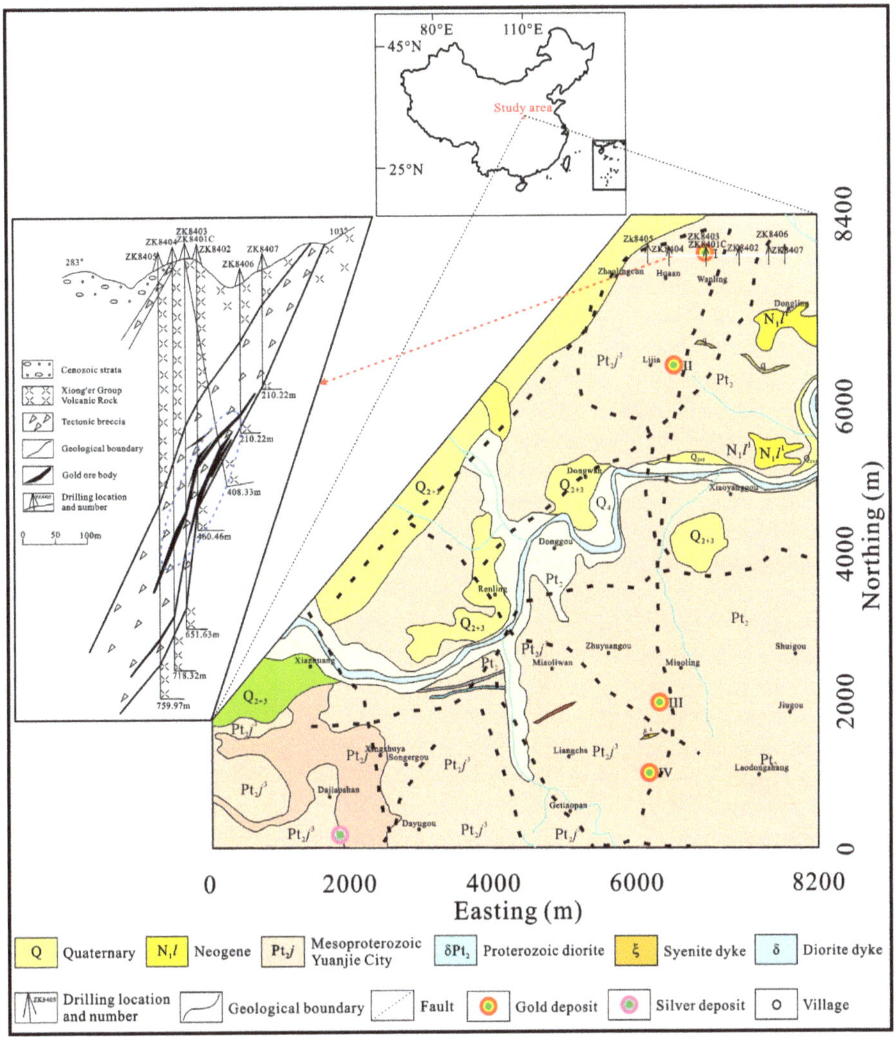

Figure 6. Map showing the geological and tectonic setting of the study area.

In 2019, a comprehensive 1:10,000 low-altitude aeromagnetic survey was conducted in the study area using a multi-rotor unmanned aerial vehicle (M600). The survey maintained an average flight altitude of 96.7 m above ground level, employing an optical pump magnetometer (Ru/GSMP-35A) as the primary aeromagnetic surveying instrument. The geomagnetic field parameters during the survey were characterized by an inclination of 52.5° and a declination of −4.5°.

The acquired total-field magnetic anomaly ($\boldsymbol{d}_{\Delta T}$) and three-component magnetic gradient anomalies ($\boldsymbol{d}_{\Delta T_x}$, $\boldsymbol{d}_{\Delta T_y}$, and $\boldsymbol{d}_{\Delta T_z}$) are presented in Figure 8a–d. The observed anomalies exhibit a distinct banded distribution pattern, which aligns closely with the regional fault structure orientation, demonstrating a strong structural control on the magnetic signature of the study area.

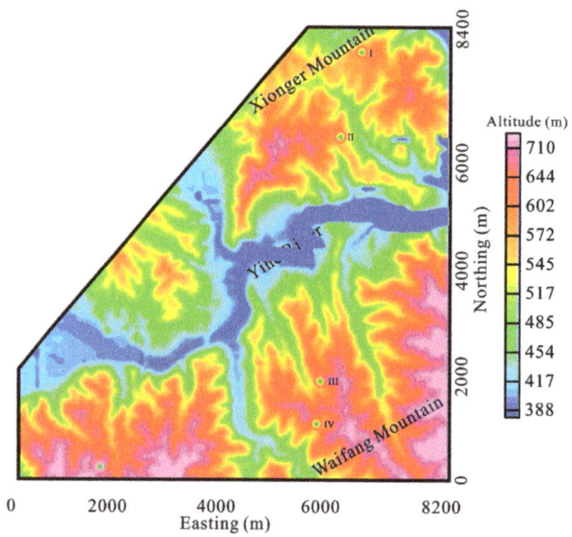

Figure 7. Topographic map of the study area.

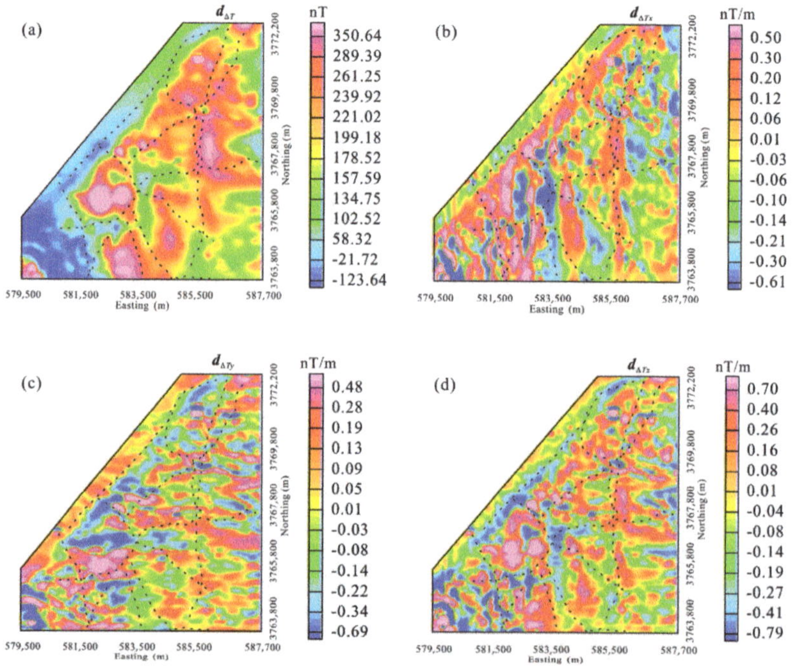

Figure 8. (**a**) Total-field magnetic anomaly. Magnetic gradient anomaly in (**b**) x-direction, (**c**) y-direction, and (**d**) z-direction.

To identify potential locations of undiscovered gold deposits, we applied the SSC method to invert the magnetic susceptibility structure of the study area. For comparative analysis, conventional joint inversion of total-field and gradient magnetic data was also performed, with both methods evaluated based on their ability to resolve known gold deposits. The inversion results are presented in Figure 9a,b, with the locations of four existing gold deposits (I–IV) annotated for reference. Using rock magnetic susceptibility data from known gold deposits, we established a cut-off threshold of 0.01 SI to delineate highly magnetic bodies (primarily andesite) in the subsurface. The conventional joint inversion results (Figure 9a) showed good correspondence with gold deposit I but failed to accurately locate gold deposit II. While gold deposits III and IV were partially resolved, their horizontal positions significantly deviated from actual locations. In contrast, the SSC-based joint inversion (Figure 9b) demonstrated superior performance, with recovered magnetic susceptibility distributions showing strong spatial correlation with all four gold deposits. These results indicate that the SSC method provides more reliable inversion outcomes, making it particularly suitable for delineating potential ore-forming areas in the study region.

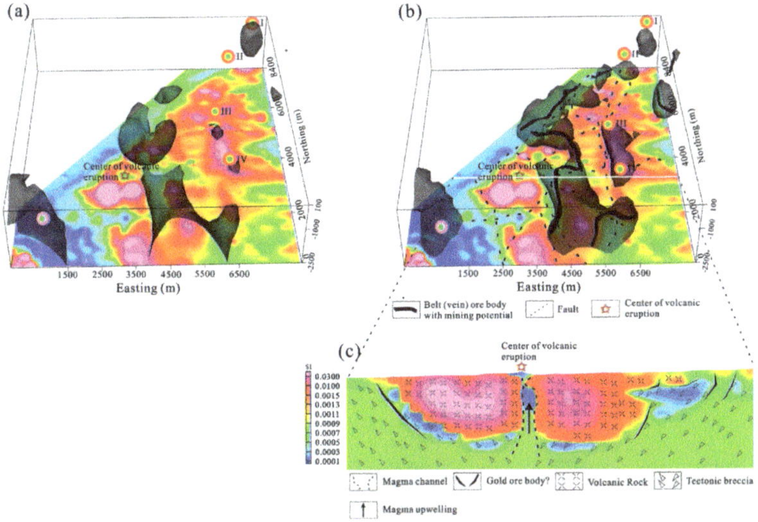

Figure 9. 3D inversion result of total-field and gradient magnetic data by the (**a**) conventional joint inversion method (I–IV are the locations of four existing gold deposits) and (**b**) SSC method. (**c**) Slice of the 3D inversion result by the SSC method after volcanic zone subdivision and the corresponding metallogenic model.

The discovered gold deposits are structurally controlled, occurring within fault-related alteration zones and exhibiting a distribution pattern strongly influenced by the basement fault tectonic belt. Based on the SSC inversion results and the fault distribution in the study area, we have identified six potential ore bodies exhibiting banded (vein-type) distributions (Figure 9b). These ore bodies predominantly strike in NE and NW directions, with vertical extents ranging from approximately 300 m to 2000 m in depth.

Our analysis revealed the presence of volcanic zones in the western portion of the study area, characterized by distinct closed-ring magnetic anomalies in the total-field data and corresponding topographic highs. Figure 9c presents the magnetic susceptibility results following volcanic zone subdivision, along with the proposed metallogenic model. We interpret the deposits in the study area as typical magmatic–hydrothermal type mineralization. During the late stages of volcanic activity, intense exhalative and hydrothermal processes transported significant quantities of metallic compounds in gaseous and liquid phases. As magma ascended along regional structural pathways, these ore-bearing hydrothermal fluids interacted with surrounding rocks under specific geological and physicochemical conditions. This process facilitated the concentration and precipitation of gold from the hydrothermal fluids, ultimately forming vein-type magmatic–hydrothermal gold deposits. Such deposit types typically exhibit substantial mining potential due to their structural control and mineralization intensity.

5. Conclusions

The proposed self-structural constraint (SSC) method, integrating total-field and gradient magnetic data through unstructured tetrahedral meshing, significantly enhances the resolution of subsurface magnetic susceptibility imaging for mineral exploration. By leveraging gradient structures of multi-component inversion results as mutual constraints, the SSC framework overcomes limitations of conventional methods in resolving complex geological settings. The derived tetrahedral magnetic gradient formulas based on Poisson's theory ensure computational accuracy, validated by synthetic hexahedral models. Theoretical tests demonstrate that SSC improves spatial resolution by 40%–60% and robustly locates magnetic sources under noise-contaminated scenarios.

Applied to a gold-bearing district in Henan Province, China, the SSC inversion delineated six concealed vein-type orebodies (300–2000 m depth) along NE- and NW-trending faults, aligning with high magnetic anomalies at andesite boundaries. Integration with the terrain and volcanic facies analysis revealed that NE-oriented faults acted as conduits for magma upwelling, supplying both metallogenic materials and structural traps for gold mineralization. These findings not only refine the ore-controlling fault model but also identify two high-priority drill targets. The SSC method bridges high-resolution geophysical imaging with critical metallogenic processes, offering a transformative approach for detecting deep-seated, structurally controlled mineral systems. This advancement underscores the pivotal role of innovative magnetic techniques in unlocking concealed resources and optimizing exploration efficiency.

Author Contributions: Conceptualization, J.C. and G.M.; methodology, J.C. and G.M.; software, J.C.; validation, J.C.; formal analysis, J.C.; data curation, J.C.; writing—original draft preparation, J.C.; writing—review and editing, J.C. and G.M.; visualization, G.M.; supervision, G.M.; project administration, J.C. and G.M.; funding acquisition, J.C. and G.M. All authors have contributed significantly and have participated sufficiently to take responsibility for this research. All authors have read and agreed to the published version of the manuscript.

Funding: This research was funded by the Excellent Young Scientist Foundation of Hunan Provincial Education Department (24B0642), and the Doctoral Startup Project of Hunan University of Arts and Science (24BSQD21).

Data Availability Statement: The original contributions presented in this study are included in the article. Further inquiries can be directed to the corresponding author(s).

Conflicts of Interest: The authors declare that they have no conflict of interest.

Appendix A

Based on Poisson's theory, the forward analytical expression of the total-field magnetic anomaly ($\boldsymbol{d}_{\Delta T}$) can be obtained following six independent analytical expressions of the second-order derivative of the gravity potential, and for each tetrahedron cell

$$\begin{aligned} V_{xx} &= \sum_{i=1}^{4}\sum_{j=1}^{3} \sin\phi\cos\theta \cdot \left\{ (-\sin\theta\cos\psi - \cos\theta\cos\phi\sin\psi)\ln\left[\xi + (\xi^2+\eta^2+Z^2)^{\frac{1}{2}}\right] \right. \\ &\left. + \cos\theta\sin\phi \cdot \tan^{-1}, \frac{-\xi\eta + (\eta^2+Z^2)\tan\psi}{Z\cdot(\xi^2+\eta^2+Z^2)^{\frac{1}{2}}} \right\}_{\xi_j}^{\xi_{j+1}} \end{aligned} \tag{A1}$$

$$\begin{aligned} V_{yy} &= \sum_{i=1}^{4}\sum_{j=1}^{3} \sin\phi\sin\theta \cdot \left\{ (\cos\theta\cos\psi - \sin\theta\sin\phi\sin\psi)\ln\left[\xi + (\xi^2+\eta^2+Z^2)^{\frac{1}{2}}\right] \right. \\ &\left. + \sin\theta\sin\phi \cdot \tan^{-1} \frac{-\xi\eta + (\eta^2+Z^2)\tan\psi}{Z\cdot(\xi^2+\eta^2+Z^2)^{\frac{1}{2}}} \right\}_{\xi_j}^{\xi_{j+1}} \end{aligned} \tag{A2}$$

$$\begin{aligned} V_{zz} &= \sum_{i=1}^{4}\sum_{j=1}^{3} \cos\phi \cdot \left\{ \sin\phi\sin\psi \cdot \ln\left[\xi + (\xi^2+\eta^2+Z^2)^{\frac{1}{2}}\right] \right. \\ &\left. + \cos\phi \cdot \tan^{-1} \frac{-\xi\eta + (\eta^2+Z^2)\tan\psi}{Z\cdot(\xi^2+\eta^2+Z^2)^{\frac{1}{2}}} \right\}_{\xi_j}^{\xi_{j+1}} \end{aligned} \tag{A3}$$

$$\begin{aligned} V_{xy} = V_{yx} &= \sum_{i=1}^{4}\sum_{j=1}^{3} \sin\phi\sin\theta \cdot \left\{ (-\sin\theta\cos\psi - \cos\theta\cos\phi\sin\psi)\cdot\ln\left[\xi + (\xi^2+\eta^2+Z^2)^{\frac{1}{2}}\right] \right. \\ &\left. + \cos\theta\sin\phi \cdot \tan^{-1}, \frac{-\xi\eta + (\eta^2+Z^2)\tan\psi}{Z\cdot(\xi^2+\eta^2+Z^2)^{\frac{1}{2}}} \right\}_{\xi_j}^{\xi_{j+1}} \end{aligned} \tag{A4}$$

$$\begin{aligned} V_{xz} = V_{zx} &= \sum_{i=1}^{4}\sum_{j=1}^{3} \sin\phi\cos\theta \cdot \left\{ \sin\phi\sin\psi \cdot \ln\left[\xi + (\xi^2+\eta^2+Z^2)^{\frac{1}{2}}\right] \right. \\ &\left. + \cos\phi \cdot \tan^{-1} \frac{-\xi\eta + (\eta^2+Z^2)\tan\psi}{Z\cdot(\xi^2+\eta^2+Z^2)^{\frac{1}{2}}} \right\}_{\xi_j}^{\xi_{j+1}} \end{aligned} \tag{A5}$$

$$\begin{aligned} V_{yz} = V_{zy} &= \sum_{i=1}^{4}\sum_{j=1}^{3} \cos\phi \cdot \left\{ (\cos\theta\cos\psi - \sin\theta\cos\phi\sin\psi)\ln\left[\xi + (\xi^2+\eta^2+Z^2)^{\frac{1}{2}}\right] \right. \\ &\left. + \sin\theta\sin\phi \cdot \tan^{-1} \frac{-\xi\eta + (\eta^2+Z^2)\tan\psi}{Z\cdot(\xi^2+\eta^2+Z^2)^{\frac{1}{2}}} \right\}_{\xi_j}^{\xi_{j+1}} \end{aligned} \tag{A6}$$

where $0 \le \theta < 2\pi$ and $0 \le \phi \le \pi$. The original coordinate system (x, y, z) was transformed from the volume integral to the surface integral by two coordinate rotations to obtain the coordinate system (X, Y) for this study. Z is constant, and the transformation process is

$$\begin{bmatrix} X \\ Y \\ Z \end{bmatrix} = \begin{bmatrix} \cos\varphi & 0 & -\sin\varphi \\ 0 & 1 & 0 \\ \sin\varphi & 0 & \cos\varphi \end{bmatrix} \begin{bmatrix} \cos\theta & \sin\theta & 0 \\ -\sin\theta & \cos\theta & 0 \\ 0 & 0 & 1 \end{bmatrix} \begin{bmatrix} x \\ y \\ z \end{bmatrix} \tag{A7}$$

To facilitate the calculation, the coordinate system (X, Y) was transformed from the surface integral to the line integral by one coordinate rotation ($0 \leq \psi < 2\pi$), and a new coordinate system (ξ, η) was obtained. The transformation process is as follows:

$$\begin{bmatrix} \xi \\ \eta \end{bmatrix} = \begin{bmatrix} \cos\psi & \sin\psi \\ -\sin\psi & \cos\psi \end{bmatrix} \tag{A8}$$

Appendix B

To obtain the forward analytical expression of the magnetic gradient anomalies in three directions ($\boldsymbol{d}_{\Delta T_x}, \boldsymbol{d}_{\Delta T_y}, \boldsymbol{d}_{\Delta T_z}$), we derived the analytical expression of the third-order derivative of gravity potential, and for each tetrahedron cell

$$\begin{aligned} \boldsymbol{V}_{xyz} = \sum_{i=1}^{4}\sum_{j=1}^{3} \Bigg\{ & \frac{A\cdot\cos\phi\cdot\left(-\sin\theta\cdot\left(\cos\psi\cdot\eta+\sin\psi\cdot\left(\xi+\sqrt{Z^2+\eta^2+\xi^2}\right)\right)\right)}{C} + \\ & \frac{A\cdot\cos\phi\cdot\left(\cos\theta\cdot\left(\sin\phi\cdot Z+\cos\phi\cdot\left(-\sin\psi\cdot\eta+\cos\psi\cdot\left(\xi+\sqrt{Z^2+\eta^2+\xi^2}\right)\right)\right)\right)}{C} + \\ & \frac{B\cdot\cos\phi\cdot\cos\psi\cdot(\cos\theta\cdot(-\eta\cdot(\cos\phi\cdot Z+\sin\phi\cdot\sin\psi\cdot\eta)))\cdot(Z^2+\eta^2)}{D} + \\ & \frac{B\cdot\cos\phi\cdot\cos\psi\cdot(\cos\psi\cdot\eta\cdot(-\cos\phi\cdot\sin\psi\cdot Z\cdot\eta+\sin\phi\cdot(2\cdot Z^2+\eta^2))\cdot\xi)}{D} + \\ & \frac{B\cdot\cos\phi\cdot\cos\psi\cdot(\sin\psi\cdot(-2\cdot\cos\phi\cdot\sin\psi\cdot Z\cdot\eta+\sin\phi\cdot(Z-\eta)\cdot(Z+\eta))\cdot\xi^2)}{D} + \\ & \frac{B\cdot\cos\phi\cdot\cos\psi\cdot(\cos\psi\cdot(\cos\phi\cdot\sin\psi\cdot Z+\sin\phi\cdot\eta)\cdot\xi^3)}{D} + \\ & \frac{B\cdot\cos\phi\cdot\cos\psi\cdot(\sin\theta\cdot Z\cdot\xi\cdot(\sin^2\psi\cdot(Z^2+\eta^2)-2\cdot\cos\psi\cdot\sin\psi\cdot\eta\cdot\xi+\cos^2\psi\cdot(Z^2+\xi^2)))}{D} \Bigg\}_{\xi_j}^{\xi_{j+1}} \end{aligned} \tag{A9}$$

$$\begin{aligned} \boldsymbol{V}_{zzy} = \sum_{i=1}^{4}\sum_{j=1}^{3} \Bigg\{ & \frac{A\cdot\cos\phi\cdot\left(\cos\phi\cdot Z+\sin\phi\cdot\left(\sin\psi\cdot\eta-\cos\psi\cdot\left(\xi+\sqrt{Z^2+\eta^2+\xi^2}\right)\right)\right)}{C} + \\ & \frac{B\cdot\cos\phi\cdot\cos\psi\cdot(\eta\cdot(\sin\phi\cdot Z-\cos\phi\cdot\sin\psi\cdot\eta)\cdot(Z^2+\eta^2))}{D} + \\ & \frac{B\cdot\cos\phi\cdot\cos\psi\cdot(\cos\psi\cdot\eta\cdot(\sin\phi\cdot\sin\psi\cdot Z\cdot\eta+\cos\phi\cdot(2\cdot Z^2+\eta^2))\cdot\xi)}{D} + \\ & \frac{B\cdot\cos\phi\cdot\cos\psi\cdot(\sin\psi\cdot(2\cdot\sin\phi\cdot\sin\psi\cdot Z\cdot\eta+\cos\phi\cdot(Z-\eta)\cdot(Z+\eta))\cdot\xi^2)}{D} + \\ & \frac{B\cdot\cos\phi\cdot\cos\psi\cdot(\cos\psi\cdot(-\sin\phi\cdot\sin\psi\cdot Z+\cos\phi\cdot\eta)\cdot\xi^3)}{D} \Bigg\}_{\xi_j}^{\xi_{j+1}} \end{aligned} \tag{A10}$$

$$V_{zzz} = \sum_{i=1}^{4}\sum_{j=1}^{3}\left\{\frac{A\cdot\cos\phi\cdot(\cos\phi\cdot Z+\sin\phi\cdot(\sin\psi\cdot\eta-\cos\psi\cdot(\xi+\sqrt{Z^2+\eta^2+\xi^2})))}{C}+\right.$$
$$\frac{B\cdot\cos\phi\cdot\cos\psi\cdot(\eta\cdot\sin\phi\cdot Z-\cos\phi\cdot\sin\psi\cdot\eta)\cdot(Z^2+\eta^2)}{D}+$$
$$\frac{B\cdot\cos\phi\cdot\cos\psi\cdot(\cos\psi\cdot\eta\cdot(\sin\phi\cdot\sin\psi\cdot Z\cdot\eta+\cos\phi\cdot(2\cdot Z^2+\eta^2))\cdot\xi)}{D}+ \quad \text{(A11)}$$
$$\frac{B\cdot\cos\phi\cdot\cos\psi\cdot(\sin\psi\cdot(2\cdot\sin\phi\cdot\sin\psi\cdot Z\cdot\eta+\cos\phi\cdot(Z-\eta)\cdot(Z+\eta))\cdot\xi^2)}{D}+$$
$$\left.\frac{B\cdot\cos\phi\cdot\cos\psi\cdot(\cos\psi\cdot(-\sin\phi\cdot\sin\psi\cdot Z+\cos\phi\cdot\eta)\cdot\xi^2)}{D}\right\}_{\xi_j}^{\xi_{j+1}}$$

where $A = \sin\phi\cdot\sin\psi$, $B = \cos\phi\cdot\cos\psi$, $C = Z^2+\eta^2+\xi\cdot(\xi+\sqrt{Z^2+\eta^2+\xi^2})$, $D = (Z^2+\eta^2)\cdot\sqrt{Z^2+\eta^2+\xi^2}\cdot(\sin^2\psi\cdot(Z^2+\eta^2)-2\cdot\sin\psi\cdot\cos\psi\cdot\eta\cdot\xi+\cos^2\psi\cdot(Z^2+\xi^2))$. $\cos\theta = -\frac{S_{yx}}{\sqrt{S_{yz}^2+S_{zx}^2}}$, $\sin\theta = -\frac{S_{zx}}{\sqrt{S_{yz}^2+S_{zx}^2}}$, $\cos\phi = -\frac{S_{yx}}{\sqrt{S_{yz}^2+S_{zx}^2+S_{xy}^2}}$, $\sin\phi = \sqrt{\frac{S_{yz}^2+S_{zx}^2}{S_{yz}^2 S_{zx}^2 S_{xy}^2}}$. S_{yz}, S_{zx}, and S_{xy} are twice the projected areas of the spatial triangle in the $y-z$, $z-x$, and $x-y$ coordinate planes, respectively. $\cos\psi = \frac{(X_{j+1}-X_j)}{\sqrt{(X_{j+1}-X_j)^2+(Y_{j+1}-Y_j)^2}}$, $\sin\psi = \frac{(Y_{j+1}-Y_j)}{\sqrt{(X_{j+1}-X_j)^2+(Y_{j+1}-Y_j)^2}}$, j represents the j th corner of the triangle, and $j+1$ represents the $j+1$ th corner of the triangle.

V_{zzy} can be transformed into V_{zzx}, V_{xxy}, V_{xxz}, V_{yyx}, and V_{yyz} via coordinate transformation. The transformation process is as follows:

$$V_{zzx}(x,y,z) = V_{zzy}(-y,x,z) \quad \text{(A12)}$$

$$V_{xxy}(x,y,z) = V_{zzy}(-z,y,x) \quad \text{(A13)}$$

$$V_{xxz}(x,y,z) = V_{zzy}(y,z,x) \quad \text{(A14)}$$

$$V_{yyx}(x,y,z) = V_{zzy}(z,x,y) \quad \text{(A15)}$$

$$V_{yyz}(x,y,z) = V_{zzy}(-x,z,y) \quad \text{(A16)}$$

V_{zzz} can be transformed into V_{xxx} and V_{yyy} with coordinate transformation. The transformation processes are

$$V_{xxx}(x,y,z) = V_{zzz}(-z,y,x) \quad \text{(A17)}$$

$$V_{yyy}(x,y,z) = V_{zzz}(x,-z,y) \quad \text{(A18)}$$

References

1. Gao, X.H.; Xiong, S.Q.; Yu, C.C.; Zhang, D.H.; Wu, C.P. The estimation of magnetite prospective resources based on aeromagnetic data: A case study of Qihe Area, Shandong Province, China. *Remote Sens.* **2021**, *13*, 1216. [CrossRef]
2. Wasilewski, P.; Fountain, D.W. The Ivrea Zone as a model for the distribution of magnetization in the continental crust. *Geophys. Res. Lett.* **1982**, *9*, 333–336.
3. Schnetzler, C.C. An estimation of continental crust magnetization and susceptibility from Magsat data for the conterminous United States. *J. Geophys. Res. Solid Earth* **1985**, *90*, 2617–2620.

4. Woodward, D.J. Inversion of aeromagnetic data using digital terrain models. *Geophysics* **1993**, *58*, 645–652.
5. Batista-Rodríguez, J.A.; Pérez-Flores, M.A.; Quiroga-Goode, G.; Gallardo, L.A. Geometry of ophiolites in eastern Cuba from 3D inversion of aeromagnetic data, constrained by surface geology. *Geophysics* **2007**, *72*, B81–B91.
6. Liu, S.; Hu, X.Y.; Liu, T.Y.; Feng, J.; Gao, W.L.; Qiu, L.Q. Magnetization vector imaging for borehole magnetic data based on magnitude magnetic anomaly. *Geophysics* **2013**, *78*, D429–D444. [CrossRef]
7. Anderson, E.D.; Zhou, W.; Li, Y.G.; Hitzman, M.W.; Monecke, T.; Lang, J.R.; Kelley, K.D. Three-dimensional distribution of igneous rocks near the Pebble porphyry Cu-Au-Mo deposit in southwestern Alaska: Constraints from regional-scale aeromagnetic data. *Geophysics* **2014**, *79*, B63–B79.
8. Li, Y.; Oldenburg, D.W. 3-D inversion of magnetic data. *Geophysics* **1996**, *61*, 394–408. [CrossRef]
9. Pilkington, M. 3-D magnetic imaging using conjugate gradients. *Geophysics* **1997**, *62*, 1132–1142.
10. Li, Y.; Oldenburg, D.W. Fast inversion of large-scale magnetic data using wavelet transforms and a logarithmic barrier method. *Geophys. J. Int.* **2003**, *152*, 251–265. [CrossRef]
11. Pilkington, M. 3D magnetic data-space inversion with sparseness constraints. *Geophysics* **2009**, *74*, L7–L15.
12. Kirkby, A.; Duan, J.M. Crustal structure of the Eastern Arunta Region, Central Australia, from magnetotelluric, seismic, and magnetic data. *J. Geophys. Res.* **2019**, *124*, 9395–9414.
13. Wigh, M.D.; Hansen, T.M.; Dossing, A. Inference of unexploded ordnance (UXO) by probabilistic inversion of magnetic data. *Geophys. J. Int.* **2020**, *220*, 37–58.
14. Davis, K.; Li, Y. Joint processing of total-field and gradient magnetic data. *Explor. Geophys.* **2011**, *42*, 199–206. [CrossRef]
15. Li, Y.; Oldenburg, D.W. Joint inversion of surface and three-component borehole magnetic data. *Geophysics* **2000**, *65*, 540–552.
16. Li, W.; Qian, J.; Li, Y. Joint inversion of surface and borehole magnetic data: A level-set approach. *Geophysics* **2020**, *85*, J15–J33. [CrossRef]
17. Liu, S.; Hu, X.; Xi, Y.; Liu, T. 2D inverse modeling for potential fields on rugged observation surface using constrained Delaunay triangulation. *Comput. Geosci.* **2015**, *76*, 18–30.
18. Abedi, M. A focused and constrained 2D inversion of potential field geophysical data through Delaunay triangulation, a case study for iron-bearing targeting at the Shavaz deposit in Iran. *Phys. Earth Planet. Inter.* **2020**, *309*, 106604.
19. Zuo, B.; Hu, X.; Wang, L.; Cai, Y.; Kass, M.A.; Leão-Santos, M.H. 3-D magnetic unstructured inversion. *J. Geophys. Res.* **2021**, *126*, e2021JB022668.
20. Paoletti, V.; Ialongo, S.; Florio, G.; Fedi, M.; Cella, F. Self-constrained inversion of potential fields. *Geophys. J. Int.* **2013**, *195*, 854–869. [CrossRef]
21. Davide, L.R.; Giovanni, F.; Luigi, F.; Simone, I.; Gabriella, C. Self-constrained inversion of microgravity data along a segment of the Irpinia fault. *J. Appl. Geophys.* **2016**, *124*, 148–154.
22. Sun, S.D.; Chen, C. A self-constrained inversion of magnetic data based on correlation method. *J. Appl. Geophys.* **2016**, *135*, 8–16. [CrossRef]
23. Vitale, A.; Maurizio, F. Self-constrained inversion of potential fields through a 3D depth weighting. *Geophysics* **2020**, *85*, G143–G156. [CrossRef]
24. Zhou, S.; Wei, Y.; Lu, P.Y.; Yu, G.R.; Wang, S.Q.; Jiao, J.; Yu, P.; Zhao, J.W. A Deep Learning Gravity Inversion Method Based on a Self-Constrained Network and Its Application. *Remote Sens.* **2024**, *16*, 995. [CrossRef]
25. Ming, Y.B.; Ma, G.Q.; Wang, T.H.; Ma, B.Z.; Meng, Q.F.; Li, Z.R. Power-Type Structural Self-Constrained Inversion Methods of Gravity and Magnetic Data. *Remote Sens.* **2024**, *16*, 681. [CrossRef]
26. Okabe, M. Analytical expressions for gravity anomalies due to homogeneous polyhedral bodies and translations into magnetic anomalies. *Geophysics* **1979**, *44*, 730–741. [CrossRef]
27. Portniaguine, O.; Zhdanov, M.S. Focusing geophysical inversion images. *Geophysics* **1999**, *64*, 874–887. [CrossRef]
28. Meng, Q.F.; Ma, G.Q.; Li, L.L.; Wang, T.H.; Han, J.T. 3-D Cross-gradient joint inversion method for gravity and gagnetic data with unstructured grids based on second-order Taylor formula: Its application to the southern Greater Khingan Range. *IEEE Trans. Geosci. Remote Sens.* **2022**, *60*, 5914816. [CrossRef]
29. Sheng, Y.; Tang, L.; Zhang, S.; Santosh, M.; Hu, X. Influence of fluid-rock interaction on gold mineralization in the Dongwan deposit, East Qinling, China: Constraints from systematic sulfur isotope and trace elements geochemistry. *Ore Geol. Rev.* **2022**, *142*, 104718. [CrossRef]

minerals

MDPI

Article

Integrating Magnetic and Self-Potential Methods for Efficient Graphite Exploration: Insights from Ji'an, Northeast China

Junchen Liu [1], Jiwei Yi [2], Chuncheng Li [3] and Ci Gao [4,*]

1 Team 608, Jilin Nonferrous Metals Geological Exploration Bureau, Changchun 130026, China
2 Team 602, Jilin Nonferrous Metals Geological Exploration Bureau, Baishan 134300, China
3 Jilin Nonferrous Metals Geological Exploration Bureau, Changchun 130026, China
4 College of Geoexploration Science and Technology, Jilin University, Changchun 130026, China
* Correspondence: gaoci22@mails.jlu.edu.cn

Abstract: Graphite, known for its exceptional electrical and thermal conductivity as well as its lubricating properties, is a critical mineral resource for various industrial applications. Due to these unique properties, graphite has gained increasing importance across various technological and industrial fields. Northeast China, rich in graphite deposits, faces significant exploration challenges owing to dense vegetation and limited bedrock exposure. This study provides a comprehensive analysis of the geological strata in the region, utilizing magnetic exploration to identify ore-forming strata beneath the Quaternary sediment. Additionally, self-potential methods were used to delineate areas of potential graphite mineralization. The effectiveness of these methods was evaluated across two well-tested and thoroughly studied geological profiles. The magnetic anomaly revealed three magnetic anomaly areas, with the low magnetic anomaly zone associated the Huangchagou Formation, indicating strong potential for graphite ore. Then, a self-potential survey was conducted across this anomalous zone, revealing a nearly east–west trending banded anomaly, suggesting the presence of graphite deposits. Trenching and drilling operations were subsequently carried out, confirming the existence of graphite ore, with total reserves exceeding 50,000 kt. This research demonstrates that the combination of magnetic surveys and self-potential methods offers a cost-effective geophysical approach for graphite exploration. These methods provide a promising strategy for discovering graphite deposits, particularly in the challenging terrains of Northeast China.

Keywords: graphite ore; magnetic method; self-potential method

1. Introduction

Graphite is a critical strategic mineral resource with extensive industrial applications and substantial economic value [1,2]. Its exceptional physical and chemical properties make it indispensable in various sectors, including metallurgy, chemistry, machinery, medical equipment, nuclear power, and the automotive and aviation industries [3–5]. Consequently, enhancing graphite resource exploration is crucial for ensuring national resource security and fostering industrial advancement [6].

Graphite deposits occur in various forms, such as stratiform, disseminated, vein-type, pegmatite type and those associated with natural iron and meteorites [7,8]. Chinese geologists further classify these deposits into regional metamorphic, contact metamorphic, and hydrothermal graphite types [9–11]. The formation and distribution of these deposits are closely linked to geological and tectonic environments. For instance, hydrothermal graphite deposits primarily form under high-pressure, high-temperature reducing conditions related to magmatic hydrothermal intrusion in carbon-bearing strata [12]. In contrast, regional metamorphic graphite deposits are typically associated with n contrast and regional structural magmatic zones such as platform uplifts and fold uplifts, which provide favorable conditions for their formation [13].

The formation mechanisms of the three types of graphite deposits mentioned above have been identified in China, with regional metamorphic graphite deposits accounting for 82% of the total resource reserves [9]. In Northeast China, the Liaoning Jilin rift belt is a significant graphite mineralization zone, situated within the Liaodong Neoproterozoic Paleozoic depression zone [14] (Figure 1a). This area represents a typical regional metamorphic graphite deposit. The most abundant deposits are primarily found in the Gaojiayu Formation of the Liaohe Group and the Huangchagou Formation of the Ji'an Rock Group, including prominent deposits in Huanren Graphite Mine [15], the Sanbanjiang deposit in Tonghua, and Shuangxing in Ji'an City [16]. The current study focuses on Ji'an City, where previous regional geological surveys have preliminarily characterized the graphite deposits within the Early Proterozoic Ji'an Group [16]. However, due to a lack of detailed geophysical investigation, understanding the spatial distribution of these deposits remains challenging, limiting the evaluation and development of graphite mineral resources in this region.

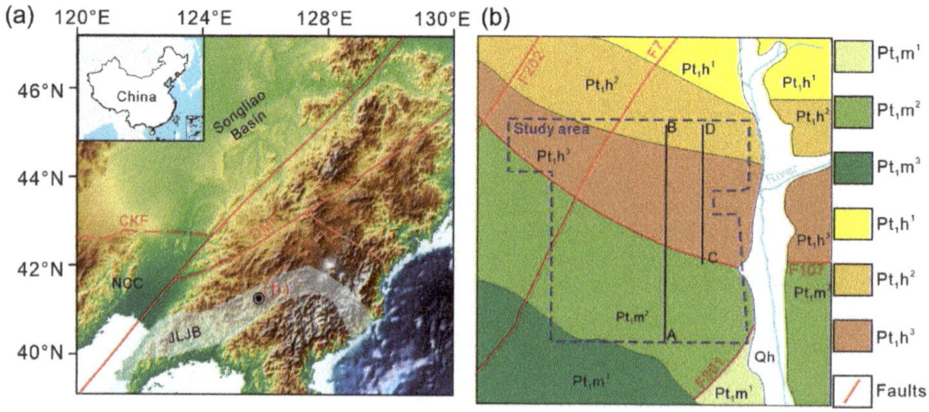

Figure 1. (**a**) Location of the study area in Northeast China. (**b**) Geological setting of study area. Red dot of the inset figure of (**a**) indicates the location in China. NCC: North China Craton; CAOB: Central Asian Orogenic Belt; CKF: Chifeng–Kaiyuan Fault; YYF: Yilan–Yitong Fault; DMF: Dunhua-Mishan Fault; JLJB: Jiao-Liao-Ji Belt; Qh: Quaternary sediment; Pt_1h^3: Section 3 of the Huangchagou Formation; Pt_1h^2: Section 2 of the Huangchagou Formation; Pt_1h^1: Section 1 of the Huangchagou Formation; Pt_1m^3: Section 3 of the Mayihe Formation; Pt_1m^2: Section 2 of the Mayihe Formation; Pt_1m^1: Section 1 of the Mayihe Formation. Red lines in (**a**,**b**) represent the faults; dashed blue zone is the magnetic investigation area. Lines A–B and C–D represent the feasibility analysis of magnetic method and self-potential method, respectively. The black triangles in (**b**) represent the measurement site of magnetic method investigation.

The layered structure and chemical bond characteristics of carbon atoms in graphite endow it with unique physical properties, such as excellent electrical conductivity [17,18]. The geo-electromagnetic method, which leverages resistivity differences in subsurface materials, has become a fundamental tool for detecting low-resistivity anomalies [19–21], especially in graphite deposits [3,22]. However, the complex genesis and variable distribution of graphite deposits mean that relying solely on a single geophysical method may be insufficient for effective exploration. To overcome these limitations, a comprehensive geophysical approach that integrates multiple methods has become essential for accurately mapping the distribution of graphite deposits [22–24].

In this study, we initially employed magnetic explorations to map the potential zone, delineating possible anomalous zones. Based on these results, a self-potential survey was conducted to further refine the delineation of graphite mineralization target areas. Induced polarization (IP) measurements in adjacent areas revealed that in sections with high graphite content, polarizabilities are unstable, making accurate measurements challenging. Additionally, the dense forest cover in the area complicates electromagnetic surveys and increases costs. In comparison to IP and electromagnetic methods, the self-potential method offers clear advantages in terms of convenience, speed, cost-effectiveness, and accuracy. These target areas were subsequently verified through drilling and trenching, leading to a preliminary assessment of the graphite resource potential in the region.

2. Geological Settings

The study area is located on the northern margin of the North China Craton and forms part of the Jiao-Liao-Ji Belt, a recognized rifting-and-collision zone [25,26]. The predominant exposed strata in this region include the Paleoproterozoic Ji'an Group, Mesozoic Jurassic formations, and Cenozoic Quaternary deposits. Of these, the Ji'an Group exhibits the most extensive distribution, comprising formations such as the Mayihe Formation (Pt_1m) and the Huangchagou Formation (Pt_1h). Two sets of U-Pb concordia ages were obtained for the Mayihe Formation. The first set yields an age of 2476 ± 22 Ma, which represents the crystallization age of Mayihe zircon remnants [27]. The second set provides an age of 2108 ± 17 Ma, indicating the crystallization age of the zircon and suggesting at the formation occurred around 2.1 Ga. The Mayihe Formation consists of a sequence of moderately metamorphosed, boron-bearing rocks, including leptite, biotite granulite, amphibolite, serpentinized marbles, and serpentinite, which collectively serve as the source rocks for non-metallic boron ores [27]. Additionally, a metamorphic zircon with an age of 1827 ± 20 Ma was found, indicating that metamorphism occurred around 1.8 Ga.

The U-Pb concordia data of Huangchagou Formation reveal two significant ages: 1838 ± 25 Ma, representing the metamorphic age of the formation, and 2144 ± 25 Ma form, corresponding to the crystallization age of the zircon. This formation is closely associated with graphite mineralization and consists of a suite of moderately metamorphosed graphite-bearing rock layers, including lithologies such as graphite-bearing granulite, graphite-bearing marble and amphibolite. The crystalline graphite within the Huangchagou Formation is primarily interpreted to have originated from organic matter that was initially trapped in sedimentary rocks and later metamorphosed under granulite facies conditions. This metamorphism occurred between 1.84 and 2.14 Ga, a period corresponding to the intense global formation of graphite [28]. As such, the regional metamorphism of the Huangchagou Formation is pivotal to graphite mineralization, serving as a significant source horizon for crystalline graphite deposits. The Mesozoic Guosong Formation is dominated by volcanic clastic rocks, whereas the Cenozoic Quaternary deposits are mainly distributed along river valleys. The basal layers throughout the study area generally exhibit an east–west strike.

The Ji'an Group strata beneath the Precambrian basement are characterized by well-developed fold and fault structures. The Sanbanjiang–Baomachuan complex syncline is a notable fold structure associated with crystalline graphite mineralization [27]. Many fault structures, particularly concentrated near the Proterozoic and Mesozoic strata in the region, indicate multiple tectonic activities. These fault structures primarily trend northwest in the earlier stages and northeast in the later stages, including the northwest-trending F107 fault zone and the northeast-trending F202, F7, and F301 fault structures.

The F107 fault zone, trending northwest, represents an early structural feature of the region characterized primarily by extensional and torsional deformation, extending over several kilometers. It occurs at the contact between the Mayihe and Huangchagou Formations but exerts minimal influence on the ore bodies. In contrast, the northeast-trending F202 and F7 fault structures are predominantly extensional, are partially obscured by the Quaternary, and intersect through the Ji'an Group, Jurassic strata, and Permian to Cretaceous granites. These faults are in the western part of the ore zone and have a deleterious impact on the integrity of the ore bodies.

3. Methods and Feasibility Analysis

3.1. Magnetic Method Exploration

Magnetic Method

Magnetic exploration is a geophysical technique that identifies subsurface geological structures and material compositions by measuring variations in the Earth's magnetic field intensity and direction [29]. This method operates on the principle that different rocks and ore bodies exhibit distinct magnetic properties, rendering it highly effective for mineral resource exploration, geological structural analysis, and oil and gas prospecting [30,31]. During magnetic exploration, the magnetic data observed represents the superposition of the Earth's normal magnetic field and an anomalous magnetic field induced by subsurface ferromagnetic materials. Consequently, the magnetic anomaly (ΔT) reflects the deviation between the actual geomagnetic field (T) and the expected normal magnetic field (T_0).

To prepare for this study, representative rock samples were collected from the surface of the mining area, and their physical properties were systematically tested. The magnetic properties of the rocks in the study area are summarized in Table 1. The magnetic parameters of different rock types in the table exhibit notable variations, with graphite-bearing granulite displaying distinctive magnetic properties. Graphite is a diamagnetic mineral, with an average magnetic susceptibility of 0.42 (10^{-6} SI) and an average remanent magnetization of 656.72 (10^{-3} A/m), both significantly lower than those of other samples. This suggests that the magnetic parameters of graphite-bearing granulite are unlikely to generate noticeable magnetic anomalies. This lower magnetic response is likely due to the presence of graphite, a non-magnetic mineral. Consequently, the magnetic anomaly of the graphite-bearing granulite in the study area is significantly weaker than that of other rocks with high-susceptibility rocks, supporting the delineation of Huangchagou and Mayihe Formations. Given the region's rugged terrain—characterized by high mountains, dense forests, thick overburden, and limited surface rock exposure—magnetic exploration is particularly well suited for delineating potential graphite-bearing zones within the Huangchagou Formation.

To optimize mineral exploration efforts and validate the effectiveness of the magnetic method in delineating the Huangchagou and Mayihe Formations, feasibility experiments were conducted along a detailed geological profile prior to the field data collection. The magnetic survey line is shown in Figure 1b, and these data were collected using a PMG-2 high-precision proton magnetometer.

The analysis of magnetic data from Profile AB in conjunction with the comprehensive geological profile reveals magnetic anomalies between sites 0 and 35, with anomaly amplitudes generally ranging from −250 nT to −600 nT. These anomalies are characterized by irregular negative patterns with significant amplitude variations (Figure 2). Based on geological profile analysis, the anomaly in the second section of the Mayihe Formation is attributed to leptite interbedded with homogeneous mixed rocks, which are widely distributed in this area. At site 21, an anomalous peak value of 0 nT was recorded, corresponding to the presence of serpentinized marble as indicated by the geological profile. From sites 13 to 21, the anomaly shows a gradual upward trend, indicating that the high anomaly at site 21 is caused by the southward dipping serpentinized marble.

Table 1. Magnetic Parameters of representative rocks in the study area.

Rock	Counts	K′ (10^{-6} SI)			Jr (10^{-3} A/m)		
		Max Value	Min Value	Average Value	Max Value	Min Value	Average Value
granite	17	1.36	0.08	0.76	2960.79	235.94	1353.73
diorite-porphyrite	8	11.47	0.01	1.57	87,751.74	185.42	9117.64
graphite-bearing granulite	8	0.93	0.05	0.42	1493.40	167.05	656.72
amphibolite	15	2.90	1.19	1.78	4122.86	2021.73	3308.77
leptite	12	1.10	0.12	0.56	1019.47	128.01	520.08
serpentinite	2	1.53	0.04	1.05	2007.53	225.82	1099.94
biotite granulite	10	1.71	0.01	0.78	32,154.67	243.10	7017.19

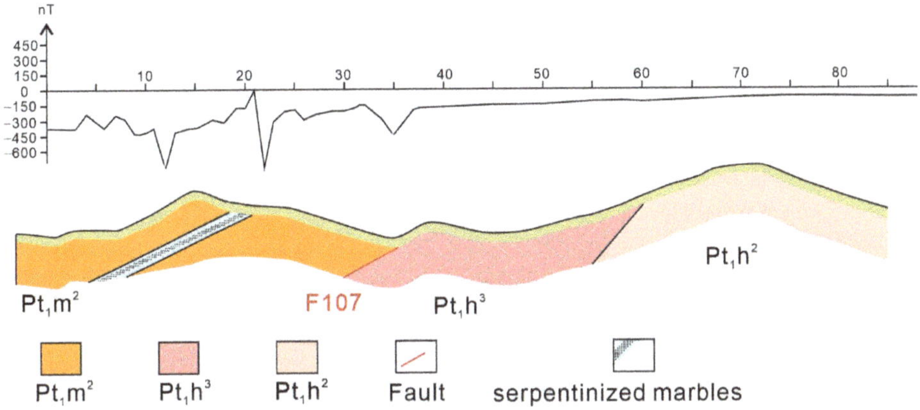

Figure 2. Magnetic anomaly data and geological mapping of Profile AB.

Near site 35, the magnetic anomaly profile reveals a distinct shift from chaotic fluctuations to a flat negative anomaly curve, which corresponds to a transition between different magnetic materials. Geological sections at this location provide found evidence of the F107 fault, which marks the contact zone between the Mayihe Formation and the Huangchagou Formation. Between sites 35 and 60, the anomaly exhibits a broad, gentle negative trend, with amplitudes generally ranging from −120 nT to −150 nT. Geological profiles indicate a substantial distribution of graphite-bearing granulite in the third section of the Huangchagou Formation, suggesting that this anomaly is due to the presence of graphite-bearing granulite. From sites 60 to 87, the anomaly is characterized by minor fluctuations in negative anomalies, with amplitudes generally ranging from −90 nT to −100 nT. Within this range, they are most prevalent from the second section of the Huangchagou Formation, indicating that these anomalies are associated with the granite.

The feasibility experiment demonstrates that magnetic measurements in the area can effectively differentiate graphite-bearing granulite from surrounding rocks within the ore zones of the third section of the Huangchagou Formation. Additionally, the method proves capable of delineating fault structures, thereby confirming its suitability for detailed geological exploration in this context.

3.2. *Self-Potential Method Exploration*

3.2.1. Self-Potential Method

The self-potential method is a passive geophysical exploration technique that detects subsurface structures by measuring self-potential differences generated by electrochemical, electrokinetic, and thermoelectric fields. The interaction between groundwater and buried ore bodies is the reason for the generation of self-potential anomalies, which are typically explained through electrochemical mechanisms and oxidation potentials [32,33]. Self-potential anomalies arise from redox reactions occurring between ore bodies (such as graphite) and/or buried metals (metallic sulfide deposits) and their surrounding rocks, groundwater, or fluids, typically manifesting as negative anomalies [34]. Due to its conductive properties, graphite especially can engage in redox reaction with surrounding oxidizers and reducers, resulting in an accumulation of excess electrons at the top of the ore body and thus generating a significant potential difference [35,36]. Previous study indicates that the negative anomaly of self-potential signal caused by the graphite could be up to several hundred millivolts [22,37]. Such a notable voltage difference makes the self-potential method a crucial technique for graphite deposit exploration [38].

3.2.2. Efficiency of the Self-Potential Method

To preliminarily assess the self-potential anomalies associated with graphite deposits, self-potential data were collected along the Profile CD using a WDJD-3A DC electrical instrument from Pentium Company. The self-potential data reveal distinct anomaly patterns across various sites (Figure 3). Between sites 0 and 5, positive anomalies were recorded with amplitudes ranging from 40 mV to 80 mV. The lithology within this segment was characterized by leptite interbedded with homogeneous mixed rocks of the second section of the Mayihe Formation, suggesting that these anomalies are likely attributed to the interbedded rock formations. In contrast, the section from sites 6 to 58 predominantly shows negative anomalies. From sites 7 to 44, anomalies range from −100 mV to −740 mV, which correlate with a substantial distribution of graphite-bearing granulite. For sites 45 to 58, anomalies range from −130 mV to −300 mV, corresponding to a widespread presence of granite. Notably, between sites 35 and 40, the anomalies range from −550 mV to −740 mV, coinciding with a known mining pit where graphite-bearing granulite protoliths are exposed.

These observations indicate that within graphite-bearing strata, the self-potential method consistently detects negative anomalies, with more pronounced negative values suggesting closer proximity to graphite deposits and higher graphite concentrations. Feasibility experiments utilizing the self-potential method further confirm that comprehensive surveys in this area can effectively delineate the distribution of graphite, accurately identify the locations of graphite deposits, and validate the reliability of this geophysical technique.

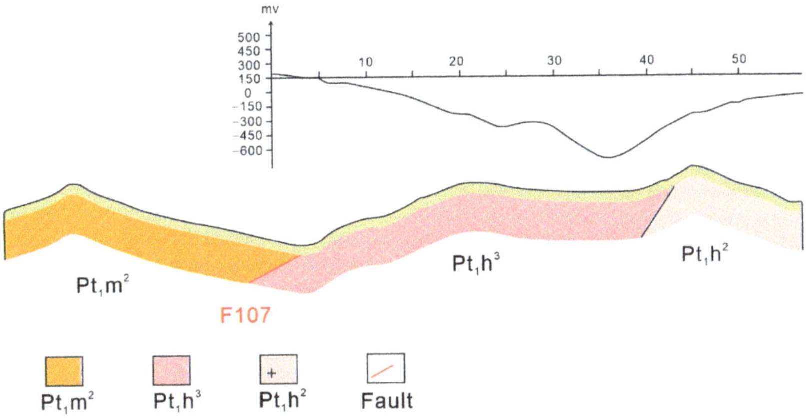

Figure 3. Self-potential anomaly data and geological mapping of the Profile CD in Figure 1b.

4. Results and Discussion

4.1. Magnetic Data

4.1.1. Magnetic Data Collection

A magnetic method investigation network was established within the study area in Figure 1b, featuring a site spacing of 20 m and a line spacing of 50 m, covering an area of approximately 3.9 km^2. Magnetic data were collected using Czech-make PMG-2 proton magnetometers. Prior to data acquisition, instrument calibration was performed, which included noise measurements, probe consistency checks, and host–instrument consistency calibration. The calibration results demonstrated good instrument consistency, confirming their suitability for field data collection. In total, 2920 measurement sites were surveyed, achieving an overall observation accuracy with a root mean square error of 2.26 nT, which exceeds the design requirement of ± 5 nT and fulfills the detection standards required for this project.

After completing the daily field data collection, we processed the raw data to obtain the total magnetic anomaly intensity. The process primarily involved the correction of the background field (normal field), diurnal variations, and topography. First, based on the location of the measurement area, we selected a relatively stable site on the aeromagnetic map to conduct 24 h diurnal observations, calculating the meaning of data with minimal fluctuations to derive the normal field for the measurement area. Additionally, we could use software to compute the International Geomagnetic Reference Field to obtain the normal field. Next, we selected a stable segment of the magnetic field within the measurement area to perform cross-section diurnal observations, which helped eliminate the influences of diurnal variations and short-period disturbances in the geomagnetic field. Finally, we carried out elevation corrections based on data from the reference station and elevation information. The aforementioned data calibration work was conducted using the cross-platform GeoIPAS V4.0 from JinweiSoft (Urimqi City in China), where we selected appropriate parameters and data to complete the calibration, ultimately obtaining the total magnetic anomaly intensity. Through these steps, we ensured the accuracy and reliability of the total magnetic anomaly intensity.

4.1.2. Characteristics of the ΔT

The magnetic field characteristics of the region are predominantly marked by a large-scale negative magnetic field background, with total magnetic anomaly intensity (ΔT) ranging from −600 nT to 200 nT. Within this framework, three primary magnetic anomalies were identified: M1, M2, and M3 (Figure 4).

The magnetic anomaly M1 exhibits values ranging from −100 nT to 20 nT and is irregularly distributed in the northern part of the survey area. This anomaly extends approximately 2000 m in length and 200–500 m in width, with a maximum anomaly value of 276 nT. While significant changes in the magnetic field gradient are observed in the northern part, likely influenced by large-scale agricultural iron wire supports and other iron objects in the area, the anomalies are generally mild elsewhere. Integrating geological mapping data with the region's magnetic properties suggests that this anomaly is associated with metamorphic rocks and granite of the second section of the Huangchagou Formation within the Ji'an Group.

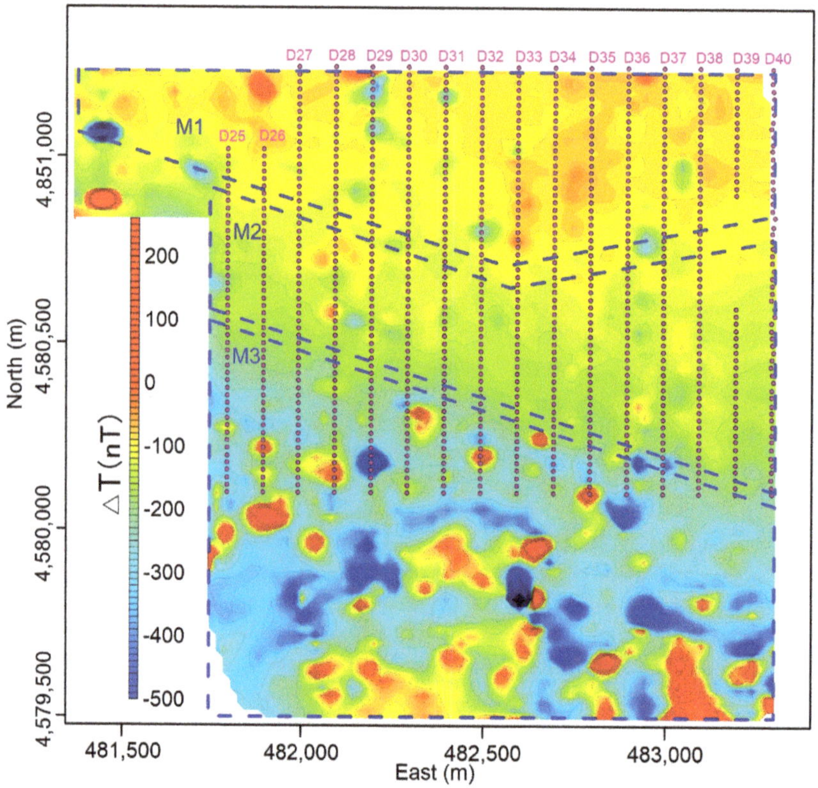

Figure 4. Magnetic anomaly in the study area. The blue dash lines and font indicate the three magnetic anomalous zones. Magenta dots and fonts indicate the self-potential sites and the survey line names, respectively.

Magnetic anomaly zone M2 ranges between −100 nT and −180 nT and is predominantly distributed in the central part of the survey area, displaying a general northwest-oriented, strip-like pattern. This anomaly spans about 2000 m in length and 500 m in width, with values varying from a maximum of −63 nT to a minimum value of −189 nT. Subsequent engineering verification, combined with tan analysis of the area's magnetic properties, suggests that the anomaly is likely caused by metamorphic rocks containing biotite and transparent pyroxene in the third section of the Huangchagou Formation of the Ji'an Group. This finding indicates that this region is a favorable location for graphite ore exploration.

The magnetic anomaly zone M3 displays a mottled distribution of high and low magnetic values. This anomaly is located at the southern part of the survey area, spanning approximately 1900 m in length and 500 to 1000 m in width, with values ranging from a minimum of −2290 nT to a maximum of 774 nT, though they generally lie between −200 nT and 300 nT. These anomalies are relatively chaotic, often exhibiting both positive and negative values with substantial gradient variations. Geological mapping and magnetic property data infer that these anomalies are related to the Mayihe Formation of the Ji'an Group and the intervening vein rocks.

In summary, the magnetic survey identified significant anomalies in the research area, correlating these variations with specific geological formations and potential mineral deposits. These findings offer valuable insights for subsequent self-potential exploration and resource assessment, enhancing the understanding of subsurface structure in the region.

4.2. *Self-Potential Data and Results*

4.2.1. Data Acquisition

The magnetic anomalous zone M2 indicates potential graphite mineralization, prompting the deployment of self-potential survey lines across this zone (Figure 4). The survey was conducted with a site spacing of 20 m and a line spacing of 50 m, covering a total of 832 sites. Data collection was carried out using the Pentium WDJD-3A DC Electrical Instrument from Pentium Technology Institute Co., Ltd. (Chongqing, China) with non-polarized electrodes. Calibration involved selecting stable electrode pairings with minimal range and measuring electrode stability before and after daily operations. The survey includes 45 inspection sites (5% of the total 832 sites), and these sites achieved an average error of 2.95 mV, exceeding the design requirement of ±3 mV.

4.2.2. Self-Potential Results

The self-potential anomalies along survey lines D36 and D38 were analyzed separately (Figure 5). Both lines exhibit significant negative anomalies with values ranging from −600 mV to −400 mV. The width of these negative anomalies varies along the survey lines, likely corresponding to the distribution of graphite-bearing strata. Previous studies have demonstrated that the self-potential anomaly values induced by graphite can exceed −400 mV [22]. Multiple locally minimum self-potential anomalies were observed on the profiles, displaying a complex anomalous structure that deviates significantly from the theoretical model [39]. These extreme negative anomalies likely reflect the self-potential characteristics of graphite-rich formations, while the variations in anomaly width may be indicative of the distribution characteristics of these strata.

The analysis results distinctly outline the boundary between the third section of the Huangchagou Formation and the second section of the Mayihe Formation (Figure 6). The third section, composed of graphite-bearing granulite, diopside granulite, and amphibolite, shows notable negative anomalies, characterized by a stable gradient in the south and significant variations in the north. In contrast, the second section of the Mayihe Formation, primarily consisting of leptite interbedded with homogeneous mixed rocks, is characterized by broad, gentle positive anomalies. These self-potential anomalies align well with the geological lithology of the survey area, confirming that the third section of the Huangchagou Formation, an essential graphite-bearing stratum, presents a −400mV anomaly with an east–west banded distribution that aligns with the known ore body. The correlation between the self-potential anomalies and geological boundaries provides a clear indication of the subsurface structure and supports the interpretation of graphite-rich formations within the survey area.

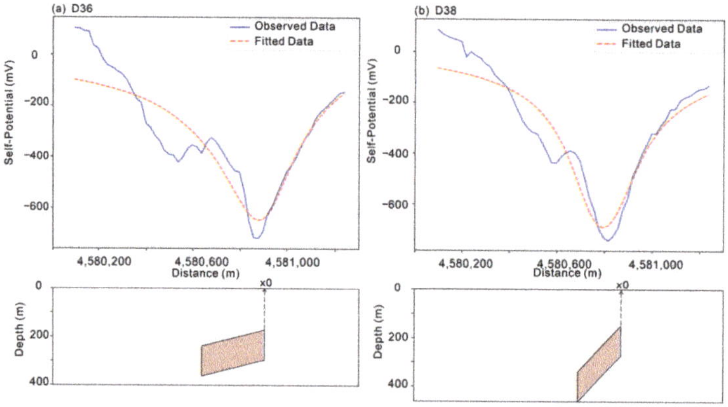

Figure 5. Self-potential anomaly of lines D36 and D38. The distance of the survey line increases from south to north. The blue lines represent the self-potential anomaly along the survey lines and the dashed red lines indicate the best inversion parameter from the particle swarm optimization algorithm. The pink color zones represent anomalies explained by inversion.

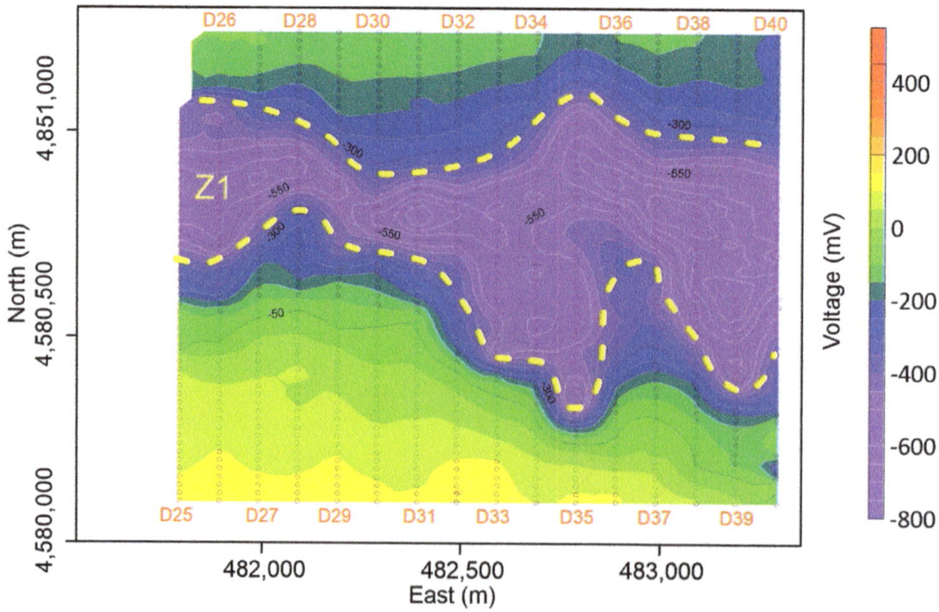

Figure 6. Self-potential anomaly. Magenta lines represent the anomaly zones Z1. Black dot lines represent the peak anomalies values. Orange lines are the designed trenches.

A significant anomaly (Z1) was identified, clearly marked by a boundary at −400 mV, delineating the third section of the Huangchagou Formation. The Z1 anomaly, with values ranging from −400 mV to −767 mV, extends as an east–west trending strip approximately

800 m wide in the east, narrowing to about 200 m in the west, with an average width of around 500 m.

To further constrain the characteristics of the self-potential anomalies observed, we conducted an inversion study using the particle swarm optimization algorithm, employing an infinitely striking sheet-like layer model. The inversion results for profiles D36 and D38 generally fit the overall shape of the self-potential anomalies; however, the finer details of the anomalies could not be fully matched. This discrepancy may reflect the complexity of the subsurface structure, data noise, and the complex terrain. Both inversions indicate that the x0 position closely aligns with the lowest data point, indicating a burial depth of approximately 150 m. The angles of the sheet-like layer vary from about 30 to 45° with a thickness over 100 m. These results imply that the graphite ore bodies are buried at a considerable depth. Therefore, further trenching and drilling are required to better constrain the distribution characteristics of the graphite deposits.

4.3. Grephite Ores Constrain from the Trenches and Drilling

To investigate the Z1 anomaly and verify the presence of deep-seated graphite ore bodies, as well as to understand their deformation characteristics, trenching and deep drilling were conducted. Six survey lines were laid out perpendicular to the strike of Z1, with nine trenches (TC1 and TC4 interconnected) and 24 drill holes positioned along these lines to verify the graphite deposits (Figure 7).

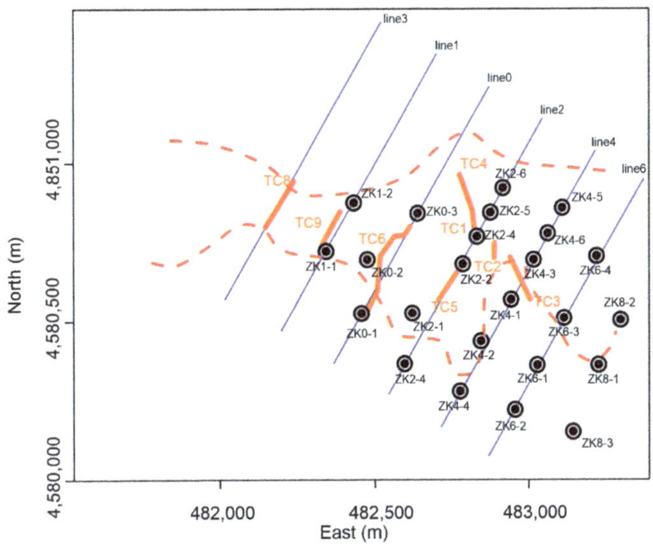

Figure 7. Distribution of the designed trenches and drill wells. The red dashed line represents the self-potential anomaly zone. Orange lines and black circles are the designed trenches and drill wells, respectively. Blue lines are the lines perpendicular to the strike of Z1.

The layout and spacing of the trenches were carefully designed to cover potential areas where graphite ore bodies may be present. Most trenches were positioned perpendicular to the strike of the ore bodies and geological boundaries, primarily to expose the ore bodies, delineate geological boundaries and structures, and facilitate the tracing and delineation of the ore body. This allowed for a better understanding of the ore bodies' shape, orientation, and thickness variations. A total of 2500 m^3 of trenching work was completed.

The trench specifications were as follows: the width of the trenches was approximately 2 m, depths ranged from 3 to 5.5 m, and bottom widths were generally between 0.8 and 1.2 m, with bedrock exposure depths from 0.5 to 1 m. All trenches met the standard requirements. Excavation results indicated that the surface layer in the study area was primarily Quaternary sediment, with an average thickness of around 3 m, underlain by bedrock. Graphite-bearing granulite bedrock was identified in all nine trenches. Geological sketches based on the trenches TC8 and TC9 results are shown in Figure 8, with graphite-bearing granulite widths of approximately 68 m in TC9 and 106 m in TC8 (Figure 8). The geological outcrops exposed in the trenches corresponded well with the geophysical anomalies, confirming the presence of potential graphite ore bodies beneath the surface.

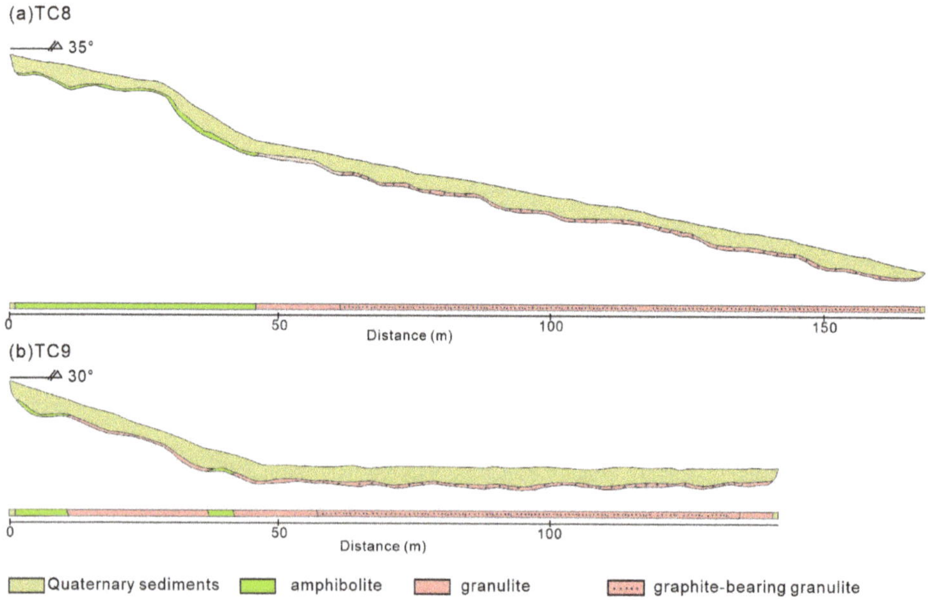

Figure 8. Geological sketch map of TC8 and TC9.

Building upon the results from trenching, a series of deep drilling operations were carried out. The drilling sites were strategically selected based on a combination of trenching data and geophysical survey results, focusing on both the center and the periphery of the Z1 anomaly as well as the graphite-ore areas revealed by trenching. This approach aimed to better define the thickness, depth, morphology, and internal structural characteristics of the ore bodies. Core samples obtained from the drilling indicate that the deep graphite ore exhibits a multi-layer distribution (Figure 9), with thickness ranging from several meters to approximately 60 m and burial depths extending from the surface to around 300 m. This reveals the substantial potential and reserves of graphite deposits in the study area.

Based on the trenching and drilling results, we further delineated two graphite ore bodies on the surface (Figure 10), well-aligned with the low-value anomaly zones identified by geophysical surveys, thereby initially validating the reliability of the geophysical anomaly data. Ore body I was controlled by two exploration trenches and several old mining pits, with the ore body trending nearly east–west and dipping to the south. The controlled length of this ore body was approximately 450 m, with an average width of 60 m.

In contrast, ore body II was controlled by five exploration trenches and old mining pits, trending northwest and dipping to the southwest with an inclination of about 20°. The controlled length of ore body II extended approximately 1100 m, with an average width of 120 m. Notably, trench TC6 revealed the widest part of the ore body, with a width of about 220 m.

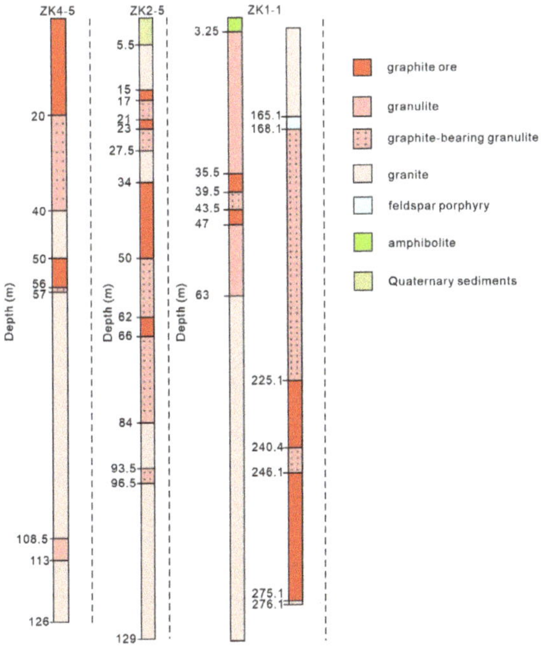

Figure 9. Borehole histogram of drillings ZK4-5, ZK2-5, and ZK1-1.

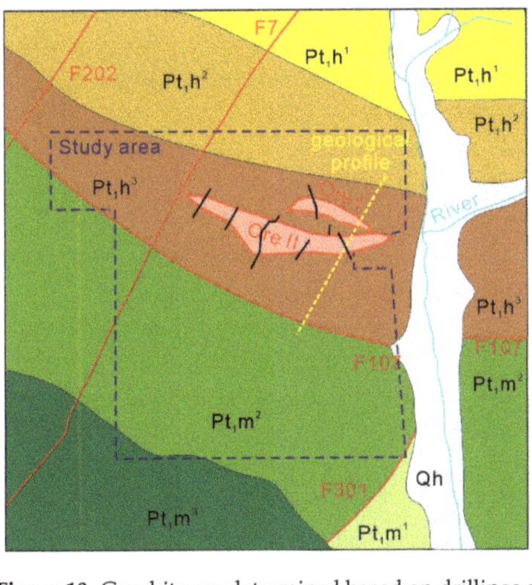

Figure 10. Graphite ore determined based on drillings and trenches. The purple polygons represent the graphite ore. The yellow dot line is profile 4 in Figure 11. The black lines on the ore bodies represent the trenches.

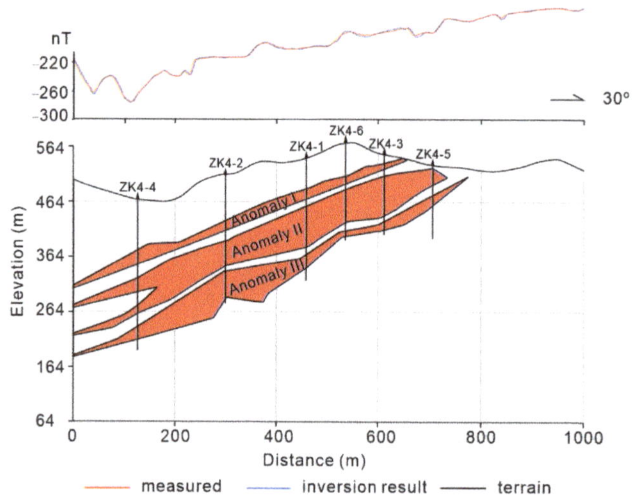

Figure 11. Magnetic 2.5D inversion result of the profile in Figure 10. Black lines represent the drillings.

4.4. Profile Characteristics of the Graphite Ore

A 2.5D magnetic inversion analysis was conducted using GeoIPAS V4.0 from JinweiSoft based on the results of trenches and drillings. This inversion program relies on the forward and inverse modeling formulas for polygonal section prisms. By assigning appropriate magnetic parameters to the theoretical model, forward calculations were performed to generate a theoretical anomaly curve. This theoretical curve was then compared with the measured anomaly curve, and through iterative adjustments of the model parameters, the theoretical curve was gradually fitted to the measured curve. This process enabled an effective interpretation and inversion of the anomaly.

During the inversion, the geological information from the trenches and drillings was set to be the prior information to construct the model. The 2.5D inversion result of line 4 crossing ore II is shown in Figure 11. This inversion was constrained by lithological information and ore depths from TC3 and drillings holes ZK4-1, ZK4-2, ZK4-3, ZK4-4, ZK4-5, and ZK4-6.

The magnetic data for line 4 yielded a fitting error of 3.5%. The result demonstrates a high level of agreement between the computational model and the observed data, thereby validating the accuracy of the model. The constrained inversion of the magnetic data, along with trench exploration and drillings data, revealed three distinct anomalies.

Anomaly I was exposed at the surface, extending 400 m with a NWW strike and a gentle southwest dip. Its magnetic susceptibility is 400*0.001A/m with a magnetic inclination of 60° and a declination of −10°. Anomaly II was located 1.4 m beneath the surface, also extending 400 m with a NWW strike and a gentle southwest dip. It has a magnetic susceptibility of 400*0.001A/m with a magnetic inclination of 59.11° and a declination of −9.98°. Anomaly III was located 7.2 m below the surface, extending 400 m with a NWW strike and a gentle southwest dip. It had a magnetic susceptibility of 400*0.001A/m with a magnetic inclination of 61° and a declination of −11°.

4.5. Graphite Reserve Assessment

The samples from ore body I from the trenches and drillings indicate an average fixed carbon grade of 3.75%. For ore body II, the drilling data reveal that the true thickness ranges from 3.76 m to 127.33 m, with an average thickness of 44.06 m. The fixed carbon grade varies from 2.00% to 9.41%, with an average of 3.27%. The reserve estimation for ore body II suggests a total ore resource of approximately 50,369 kt. This corresponds to a total mineral content of 1653 kt. These findings highlight the significant potential of ore II as a valuable resource for further development, and they will play a key role in guiding future mining and processing operations.

5. Conclusions

This study systematically evaluates the effectiveness of an integrated exploration strategy combining magnetic surveys, self-potential methods, trenches, and drillings for identifying crystalline graphite deposits in Ji'an City, Jilin Province. Initially, magnetic surveys were conducted to assist in geological mapping, leading to the identification of three significant magnetic anomaly zones (M1, M2, and M3). By analyzing the magnetic susceptibility characteristics of the rocks in the area, a potential mineralization zone was delineated, which helped focus the exploration efforts. Among these zones, the M2 anomaly zone stood out for its considerable mineralization potential. Following this, self-potential survey lines were deployed across the M2 anomaly zone, revealing a prominent east–west trending self-potential anomaly. Based on this, drilling and trenching operations were carried out, confirming that the anomaly was indeed associated with graphite deposits. The estimated reserve in this zone is potentially over 50 million tons. Given the challenging, heavily vegetated terrain of Northeast China, this study demonstrates that a combined geophysical exploration strategy utilizing both magnetic and self-potential methods offers not only offers high cost-effectiveness but also provides accurate and reliable results in the exploration of graphite deposits.

Author Contributions: Conceptualization, J.L., C.L.; data collection, processing, interpretation, validation, J.L., J.Y. and C.G.; writing—original draft preparation, J.L. and C.G., writing—review and editing, J.L., J.Y., C.L. and C.G. All authors have read and agreed to the published version of the manuscript.

Funding: This research received no external funding.

Data Availability Statement: The original contributions presented in this study are included in the article. Further inquiries can be directed to the corresponding author.

Acknowledgments: We would like to thank the three anonymous reviewers and the editor for their valuable comments and guidance. We extend our heartfelt gratitude to the personnel involved in the field data collection for their hard work and dedication.

Conflicts of Interest: The authors declare no conflict of interest.

References

1. Gautneb, H.; Rønning, J.S.; Engvik, A.K.; Henderson, I.H.C.; Larsen, B.E.; Solberg, J.K.; Ofstad, F.; Gellein, J.; Elvebakk, H.; Davidsen, B. The Graphite Occurrences of Northern Norway, a Review of Geology, Geophysics, and Resources. *Minerals* **2020**, *10*, 626. [CrossRef]
2. Jara, A.D.; Betemariam, A.; Woldetinsae, G.; Kim, J.Y. Purification, Application and Current Market Trend of Natural Graphite: A Review. *Int. J. Min. Sci. Technol.* **2019**, *29*, 671–689. [CrossRef]
3. Baranwal, V.C.; Rønning, J.S.; Larsen, B.E.; Su, Y.; Zhang, B.; Liu, Y.; Ren, X.; Gautneb, H.; Gellein, J. A Case History of Graphite Exploration in North Norway Integrating Various Geophysical Surveys. *Minerals* **2024**, *14*, 266. [CrossRef]

4. Cui, N.; Sun, L.; Bagas, L.; Xiao, K.; Xia, J. Geological Characteristics and Analysis of Known and Undiscovered Graphite Resources of China. *Ore Geol. Rev.* **2017**, *91*, 1119–1129. [CrossRef]
5. Sun, L.; Xu, C.; Xiao, K.; Zhu, Y.; Yan, L. Geological Characteristics, Metallogenic Regularities and the Exploration of Graphite Deposits in China. *China Geol.* **2018**, *1*, 424–433. [CrossRef]
6. Zhang, Y.; An, Z.; Liang, S.; Zhai, F.; Zhang, S.; Ju, N.; Jiang, P.; Xiao, R. Distribution Characteristics, Genetic Types and Prospecting Progress of Graphite Deposits. *Geol. China* **2022**, *49*, 135–150.
7. Luque, F.J.; Huizenga, J.-M.; Crespo-Feo, E.; Wada, H.; Ortega, L.; Barrenechea, J.F. Vein Graphite Deposits: Geological Settings, Origin, and Economic Significance. *Min. Depos.* **2014**, *49*, 261–277. [CrossRef]
8. Simandl, G.J.; Paradis, S.; Akam, C. Graphite Deposit Types, Their Origin, and Economic Significance. *Br. Columbia Minist. Energy Mines Br. Columbia Geol. Surv.* **2015**, *3*, 163–171.
9. Yan, L.; Gao, S.; Chen, Z.; Jiao, L.; Sun, L.; Liu, Y.; Zhou, W. Metallogenic Characteristics and Metallogenic Zoning of Graphite Deposites in China. *Geol. China* **2018**, *45*, 421–440.
10. Ren, Y.; Li, J.; Sun, X.; Li, Z.; Sun, Z. Genesis of Metal Sulfides and Its Significance on Graphite Mineralization in the Huangyangshan Graphite Deposit, East Junggar, Xinjiang Province, China. *Minerals* **2022**, *12*, 1450. [CrossRef]
11. Li, H.; Hong, T.; Liu, S.; Ke, Q.; Yang, Z.; Ma, Y.; Wang, X.; Niu, L.; Kang, K.; Xu, X. Genesis of the Graphite from the Tugeman Graphite Deposit, Xinjiang, China: Evidence for Carbon Isotope Refining by Fluids Associated with the Ductile Shear Zone. *Minerals* **2023**, *13*, 1328. [CrossRef]
12. Li, C.; Wang, D.; Zhao, H.; Pei, H.; Li, X.; Zbou, L.; Du, A.; Qu, W. Minerogenetic regularity of praphite deposites in China. *Miner. Depos.* **2015**, *34*, 1223–1236. [CrossRef]
13. Zhang, H.-F.; Zhai, M.-G.; Santosh, M.; Wang, H.-Z.; Zhao, L.; Ni, Z.-Y. Paleoproterozoic Granulites from the Xinghe Graphite Mine, North China Craton: Geology, Zircon U–Pb Geochronology and Implications for the Timing of Deformation, Mineralization and Metamorphism. *Ore Geol. Rev.* **2014**, *63*, 478–497. [CrossRef]
14. Tam, P.Y.; Zhao, G.; Liu, F.; Zhou, X.; Sun, M.; Li, S. Timing of Metamorphism in the Paleoproterozoic Jiao-Liao-Ji Belt: New SHRIMP U–Pb Zircon Dating of Granulites, Gneisses and Marbles of the Jiaobei Massif in the North China Craton. *Gondwana Res.* **2011**, *19*, 150–162. [CrossRef]
15. Wu, C.; Qu, T. Geological characteristics and genetic research on Heigou graphite ore deposit in Huanren country. *Geol. Build. Mater.* **1994**, *4*, 25–27.
16. Zhang, Q.; Liu, S. Characteristics and prospecting indicators of crystalline graphite deposits in Ji'an area, Jilin Province. *Jilin Geol.* **2014**, *33*, 60–62.
17. Murphy, B.S.; Huizenga, J.M.; Bedrosian, P.A. Graphite as an Electrically Conductive Indicator of Ancient Crustal-Scale Fluid Flow within Mineral Systems. *Earth Planet. Sci. Lett.* **2022**, *594*, 117700. [CrossRef]
18. Loukola-Ruskeeniemi, K.; Hyvönen, E.; Airo, M.-L.; Lerssi, J.; Arkimaa, H. Country-Wide Exploration for Graphite- and Sulphide-Rich Black Shales with Airborne Geophysics and Petrophysical and Geochemical Studies. *J. Geochem. Explor.* **2023**, *244*, 107123. [CrossRef]
19. Li, Y.; Weng, A.; Xu, W.; Zou, Z.; Tang, Y.; Zhou, Z.; Li, S.; Zhang, Y.; Ventura, G. Translithospheric Magma Plumbing System of Intraplate Volcanoes as Revealed by Electrical Resistivity Imaging. *Geology* **2021**, *49*, 1337–1342. [CrossRef]
20. Li, Y.; Weng, A.; Zhou, Z.; Guo, J.; Li, S.; Ventura, G.; Xu, W. Crustal Root Shapes the Plumbing System of a Monogenetic Volcanic Field as Revealed by Magnetotelluric Data. *Earth Planet. Sci. Lett.* **2024**, *626*, 118523. [CrossRef]
21. Jia, Z.; Wang, Y.; Li, Y.; Xu, C.; Wang, X.; Lu, W. Magnetotelluric Closed-Loop Inversion. *IEEE Trans. Geosci. Remote Sens.* **2023**, *61*, 5923511. [CrossRef]
22. Heritiana, A.R.; Riva, R.; Ralay, R.; Boni, R. Evaluation of Flake Graphite Ore Using Self-Potential (SP), Electrical Resistivity Tomography (ERT) and Induced Polarization (IP) Methods in East Coast of Madagascar. *J. Appl. Geophys.* **2019**, *169*, 134–141. [CrossRef]
23. Bhattacharya, B.B.; Shalivahan; Jardani, A.; Bera, A. Three-dimensional Probability Tomography of Self-potential Anomalies of Graphite and Sulphide Mineralization in Orissa and Rajasthan, India. *Near Surf. Geophys.* **2007**, *5*, 223–230. [CrossRef]
24. Liang, Y.; Xia, Q.; Zhao, M.; Bi, R.; Liu, J. Application and Significance of the Wavelet–Fractal Method on the Data of the Induced Polarization Method in the Graphite Deposits of Datong, China. *Minerals* **2023**, *13*, 760. [CrossRef]
25. Zhao, G.; Zhai, M. Lithotectonic Elements of Precambrian Basement in the North China Craton: Review and Tectonic Implications. *Gondwana Res.* **2013**, *23*, 1207–1240. [CrossRef]
26. Zhu, J.; Liu, F.; Wang, F.; Xu, W.; Liu, F.; Shi, C. Carbon Isotope and Geochemical Characteristics of the Paleoproterozoic Graphite Deposits in the Jiao-Liao-Ji Belt, North China Craton: Implications for Genesis and Depositional Environment. *Precambrian Res.* **2021**, *362*, 106320. [CrossRef]
27. Bureau of Geology and Mineral Resources of Jilin Province. *Regional Geology of Jilin Province*; Geology Press: Beijing, China, 1988.

28. Parnell, J.; Brolly, C.; Boyce, A.J. Graphite from Palaeoproterozoic Enhanced Carbon Burial, and Its Metallogenic Legacy. *Geol. Mag.* **2021**, *158*, 1711–1718. [CrossRef]
29. Jia, Z.; Li, Y.; Wang, Y.; Li, Y.; Jin, S.; Li, Y.; Lu, W. Deep Learning for 3-D Magnetic Inversion. *IEEE Trans. Geosci. Remote Sens.* **2023**, *61*, 5905410. [CrossRef]
30. Du, W.; Cheng, L.; Li, Y. L_p Norm Smooth Inversion of Magnetic Anomaly Based on Improved Adaptive Differential Evolution. *Appl. Sci.* **2021**, *11*, 1072. [CrossRef]
31. Ji, N.; Qin, X.; Du, W.; Wang, Z.; Liu, Y.; Zhang, T. The Application of Aeromagnetic Three-Dimensional Geophysical Inversion in the Exploration of Lead-Zinc Polymetallic Deposits in Forested Areas. *IEEE Geosci. Remote Sens. Lett.* **2024**, *21*, 7506505. [CrossRef]
32. Sato, M.; Mooney, H.M. The Electrochemical Mechanism of Sulfide Self-Potentials. *Geophysics* **1960**, *25*, 226–249. [CrossRef]
33. Kilty, K.T. On the Origin and Interpretation of Self-Potential Anomaly. *Geophys. Prospect.* **1984**, *32*, 51–62. [CrossRef]
34. Kontny, A.; Friedrich, G.; Behr, H.J.; de Wall, H.; Horn, E.E.; Möller, P.; Zulauf, G. Formation of Ore Minerals in Metamorphic Rocks of the German Continental Deep Drilling Site (KTB). *J. Geophys. Res.* **1997**, *102*, 18323–18336. [CrossRef]
35. Naudet, V.; Revil, A.; Bottero, J.-Y.; Bégassat, P. Relationship between Self-potential (SP) Signals and Redox Conditions in Contaminated Groundwater. *Geophys. Res. Lett.* **2003**, *30*, 21. [CrossRef]
36. Hamilton, S.M.; Hattori, K.H. Spontaneous Potential and Redox Responses over a Forest Ring. *Geophysics* **2008**, *73*, B67–B75. [CrossRef]
37. Stoll, J.; Bigalke, J.; Grabner, E.W. Electrochemical Modelling of Self-Potential Anomalies. *Surv. Geophys.* **1995**, *16*, 107–120. [CrossRef]
38. Yang, J.; Wang, Q.; Wang, X.; Ning, K.; Guo, J.; Zhou, Q.; Gao, H. Electrical characteristics and prospecting evaluation of Panzhihua crystalline graphite deposit in Sichuan Province. *Geol. China* **2022**, *49*, 284–297.
39. Biswas, A. A Review on Modeling, Inversion and Interpretation of Self-Potential in Mineral Exploration and Tracing Paleo-Shear Zones. *Ore Geol. Rev.* **2017**, *91*, 21–56. [CrossRef]

minerals

MDPI

Article

Study of the Genesis Process and Deep Prospecting Breakthrough in the Gouli Ore Concentration of the East Kunlun Metallogenic Belt Using Audio Magnetotelluric Data

Ji'en Dong [1,2], Peng Wang [1,*], Hua Li [1], Huiqing Zhang [1], Mingfu Zhao [1], Haikui Tong [1,*], Xiaoliang Yu [1], Jie Li [1] and Binshun Zhang [3]

1 Qinghai Geological Survey, MNR Technology Innovation Center for Exploration and Exploitation of Strategic Mineral Resources in Plateau Desert Region, Xining 810000, China; dje@email.cugb.edu.cn (J.D.); lihua625203@163.com (H.L.); zhhq1114@163.com (H.Z.); zmfforever@126.com (M.Z.); xiaoliangyu1988@163.com (X.Y.); 17725261391@163.com (J.L.)

2 School of Geophysics and Information Technology, China University of Geosciences, Beijing 100083, China

3 The Qinghai Third Nonferrous Metals Geological Exploration Institute, Xining 810012, China; zbs17709784463@163.com

* Correspondence: tonghaikui2024@163.com (H.T.); wangpeng20240426@163.com (P.W.)

Abstract: The East Kunlun Orogenic Belt is an essential part of the Qin-Qikun composite orogenic system, the most crucial orogenic belt in Qinghai Province, and an important gold ore-producing area in China. The Gouli gold field in its eastern section is one of the most important gold fields discovered in the belt in recent years. The Mailong mining area is an important gold mining area in the Gouli ore-concentrated area. The area has experienced frequent and intense magmatic activity, with intrusive rock bodies extensively exposed and intersected by a complex network of fault structures, providing excellent geological conditions for the formation of gold deposits. However, it is difficult to explore due to high altitude, poor transportation, and shallow coverage. This study used an audio magnetotelluric sounding method to track the deep direction and inclination of known mineral belts in the Mailong mining area, and identified mineral exploration targets, providing a basis for mineral exploration. Subsequently, a gold ore body was discovered through drilling verification, achieving a breakthrough in deep mineral exploration. The electromagnetic exploration method works well for exploring structurally altered rock-type gold deposits in plateau desert areas, and combined with the results of this electromagnetic exploration, a metallogenic geological model and genesis process of the Mailong mining area has been constructed.

Keywords: East Kunlun Orogenic Belt; Gouli ore-concentrated area; electromagnetic exploration; deep prospecting breakthrough

1. Introduction

The East Kunlun Orogenic Belt is located in the northeastern part of the Qinghai-Tibet Plateau [1], stretching from the East Kunlun to the Elashan area in an east–west direction [2]. It is an important part of the Central Orogenic Belt [3]. The East Kunlun Orogenic Belt is cut diagonally by the Altyn Tagh tectonic belt in the west, connected to the western Qinling tectonic belt in the east, covered by the Qaidam Basin in the north, and connected to the Bayankela tectonic belt in the south [4]. From north to south, it is divided into East Kunlun North Terrane, East Kunlun South Terrane, and Bayankela Terrane, bounded by two suture belts of Middle Kunlun and South Kunlun [5]. The East Kunlun orogenic belt has experienced the evolution of four different tectonic cycles: the pre-Cambrian, early Paleozoic, late Paleozoic to early Mesozoic, and late Mesozoic to Cenozoic, with widespread distribution of intrusive rocks and volcanic rocks [6,7]. It has experienced the Proto-Tethys, Paleo-Tethys, and Neo-Tethys evolutionary processes from the Paleoproterozoic to the present [8–11], forming a large number of metal deposits composed of Au-Cu-Co-Ni-Fe-Pb-Zn polymetallic belts [12]. The unique geotectonic position, complex tectonic environment, frequent magmatic activities, and different degrees of metamorphism of the East Kunlun Orogenic Belt have resulted in a variety of metal mineralization types and rich mineral resources in the belt and other geological and tectonic environments control the formation and transformation of different kinds of mineral deposits. It is considered one of the important gold mineralization regions in the Tethys tectonic domain [13] and an essential potential base for mineral resources in China. Wulonggou and Gouli gold mines have been discovered successively, earning the reputation of the "Golden Belt of Qinghai Province" [14,15].

The Gouli ore-concentrated area in the eastern section of the East Kunlun metallogenic belt is one of the most important gold fields discovered recently; gold mineralization is closely related to regional magmatic–tectonic activity [16–19]. The Gouli ore-concentrated area is mainly related to the Au-Ag-Pb-Zn mineralization series associated with the Indosinian tectonic fluids [20]. A series of large and medium-sized gold deposits are represented by Mailong, Seri, Dareer, Guoluolongwa, Asha, Annage, Walega, Kengdenongshe, Delong, Nagekangqieer, and Luotuogou, with excellent mineralization geological conditions (Figure 1b) [21–23].

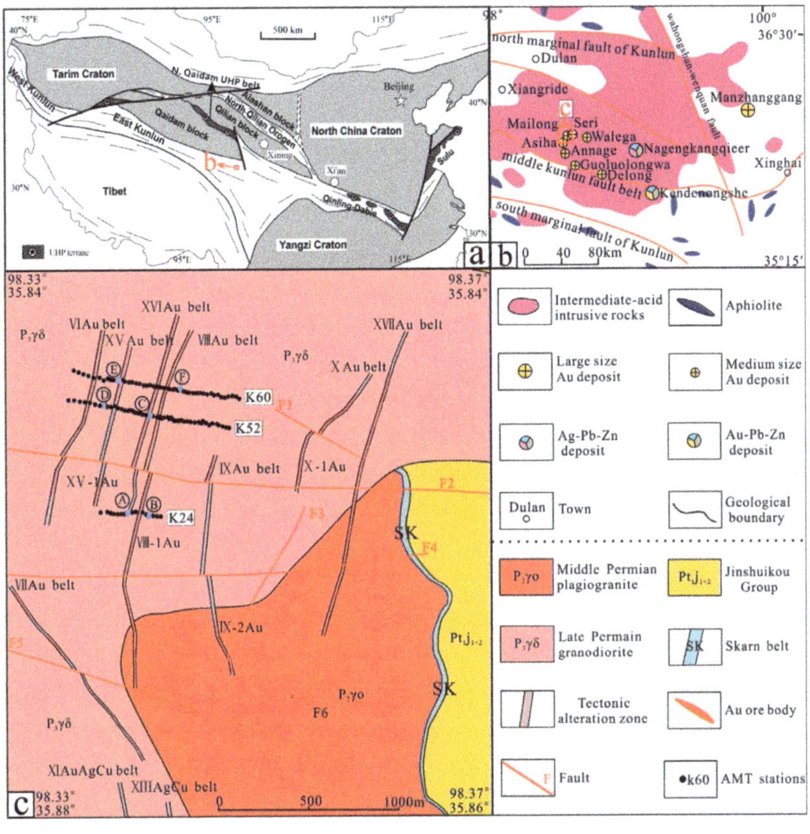

Figure 1. (**a**) Schematic maps showing major tectonic units of China [24] and the location of the study area. (**b**) Distribution map of intrusive rocks, ophiolites, and deposits in the Gouli ore-concentrated area. (**c**) Geological and mineralogical map showing major tectonic structures and AMT station locations in the Mailong gold mining area.

The Mailong mining area is one of the important mineral deposits discovered in the Gouli ore-concentrated area in recent years. However, it is located at a high altitude with complex topography and covered locations. These conditions have limited previous exploration methods. In the past, mineral exploration was mainly based on surface geological surveys and chemical anomalies, and the lack of deep mineral information slowed progress. To address this issue, AMT exploration was used to study mineral patterns controlled by tectonics and pinpoint areas for deeper exploration, resulting in significant mineral discoveries. The magnetotelluric method is currently one of the primary methods for exploration, and it is an indispensable way to study the structure of the crust and upper mantle from a conductive perspective [25,26]. At the same time, it is extensively used in connection with waste site exploration [27]. AMT observes the electrical structure within a depth range of tens of meters to several kilometers by observing the geomagnetic field signal's audio frequency band (0.1 Hz~100 kHz). Due to its fast, efficient, and cost-effective advantages, with a considerable detection depth, it is widely used in mineral resource exploration and has achieved good results [28–31].

2. Geological and Geophysical Background

The Mailong gold mining area is located north of the Middle Kunlun fault, adjacent to the Dareer and Seri gold mining areas, with several ore deposit points such as Qiangkou nan and Asia distributed in the surrounding area. Due to the influence of frequent and intense magmatic activity, mainly during the Indosinian period–Hualixi period and the North Kunlun fault structure, the area is characterized by NNW and NE-oriented fault structures dominated by compression and pressure torsion (Figure 1c) [32,33]. The extensive exposure of intrusive rocks and the crisscrossing of fault structures in the area have created excellent geological conditions for mineralization [34]. The Mailong mining area extensively exposes intrusions of medium-acidic rocks, mainly Late Permian granodiorite and Middle Permian plagiogranite. Late Permian granodiorite is the most widely distributed rock in the mining area, with most of the ore-bearing alteration zones occurring within it, making it the main rock and ore-bearing rock of the area; Middle Permian plagiogranite is exposed in the eastern part of the mining area, consisting of rock masses and rock cores, with some tectonic fracture zones occurring within it. The main strata exposed internally are the Paleoproterozoic Jinshuikou group, mainly distributed in the eastern and northern parts of the mining area [21,33]. There are multiple different faults in the mining area, including those trending nearly east–west and north–northeast [21]. The nearly east–west structures are regional structures related to early mineralization and control the distribution of strata and magmatic rocks in the area, serving as important ore-controlling structures. The north–northeast structures are secondary structures characterized by thrust faults with multiple periods of activity. These secondary structures often cut through ore bodies and are the main structures for ore guidance and hosting in the area.

Due to the varying levels of geophysical work, the study area currently has a Bouguer gravity anomaly map at a scale of 1:1,000,000 and a high-precision magnetic anomaly map at a scale of 1:50,000, which means they cover different areas and reflect different scales of fault structures (as shown in Figure 2a,b). The Bouguer gravity anomalies mainly reflect the large structural features in the region, while the magnetic anomalies primarily reflect secondary structures.

The 1:1,000,000 Bouguer gravity anomaly in the eastern section of the East Kunlun is characterized by a steeply variable gravity gradient zone with a width of 30–40 km and the main body spreading in the NWW direction, and the distribution of this zone coincides with the Middle Kunlun fault belt, which is an important tectonic unit boundary fault. The Xiangride-De Long fault is mainly characterized by a dense Bouguer gravity anomaly gradient zone trending northwest, which is an important controlling structure for mineral deposits across the Gouli mining deposit area. The Gouli mineral deposit area is located on a dense Bouguer gravity anomaly gradient zone, which trends in a northwesterly direction and exhibits the characteristic of being denser in the north and looser in the south

(Figure 2a). The widespread presence of the basement tectonic layer from the Archean to the Proterozoic is an important reason for this area's high regional gravity anomaly.

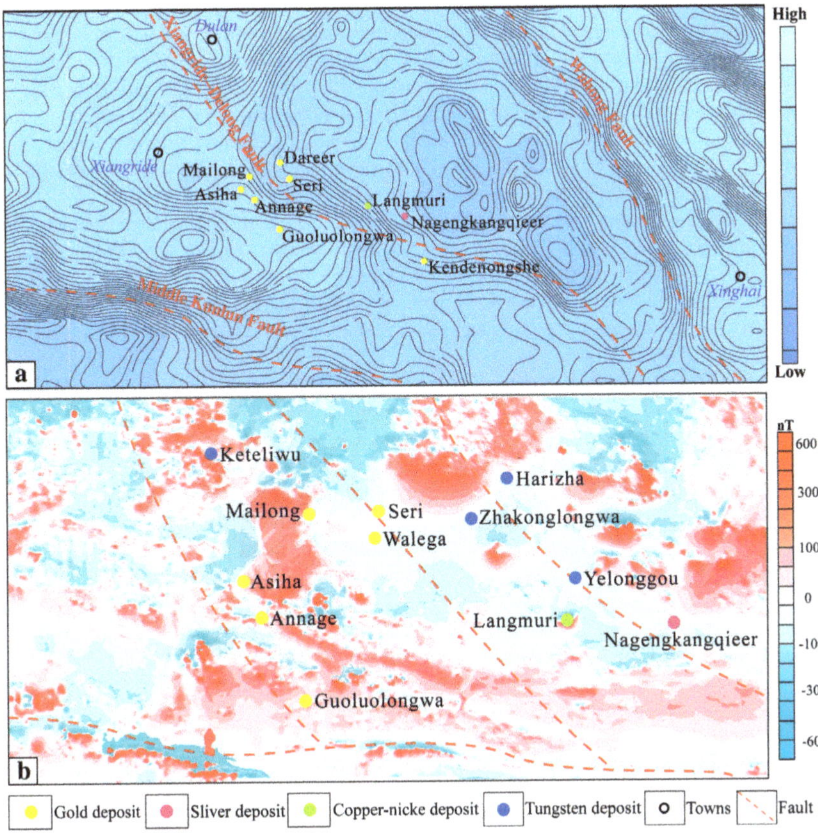

Figure 2. (**a**) The 1:1,000,000 Bouguer gravity anomaly map in the Gouli mineral deposit area. (**b**) The 1: 1:50,000 high-precision magnetic anomaly map in the Gouli mineral deposit area.

The magnetic anomaly in the research area is manifested as a high-intensity anomaly band extending from east to west. The anomaly is continuous and strong, and positive and negative, indicating that the East Kunlun region is composed of multiple geological blocks extending from east to west. Deep major faults separate these blocks, and within these blocks are a large number of highly magnetic deep metamorphic rocks or basic-ultrabasic rock bodies. The Gouli area is located at the eastern end of the high-intensity magnetic anomaly zone, with characteristics of positive and negative anomaly gradient belts.

The high-precision magnetic survey conducted in the Gouli area at a scale of 1:50,000 shows that the distribution of abnormal morphological features in the region is partitioned (Figure 2b). The intensity of magnetic anomalies is closely related to regional strata, magmatic rocks, and structures. The magnetic anomalies are distributed in bands or blocks on the plane, with extension directions consistent with the distribution of strata, magmatic rocks, and regional structural lines. In most areas, the amplitude of magnetic anomalies is low, with several high magnetic anomalies related to medium-acidic intrusive rock bodies. The discovered gold and lead–zinc ore deposits such as Mailong, Guoluolongwa, Walega, and Kengdenongshe are mainly distributed in the transition zone of positive–negative magnetic anomalies in the area near the positive magnetic anomaly belt.

3. Data Collection, Processing, and Analysis

3.1. AMT Data Acquisition and Processing

Based on the fundamental principle of EM field propagation, the penetration depth of an electromagnetic signal is directly related to the square root of its frequency. The MT technique records a series of naturally occurring electromagnetic signals on the earth's surface to investigate the resistivity structure of the lithosphere [35]. The AMT data used in this study were collected in 2023 by the Qinghai Geological Survey. The average station spacing is ~20 m and ~10 m for the focus area. Polarization is defined with true north as the positive *x*-axis and east as the positive *y*-axis. Orthogonal Ex, Ey electric field components and Hx, Hy magnetic field components were recorded at each station.

The AMT profiles (K60, K52, K24) were oriented perpendicular to the strike of the Au ore belts (the black dots in Figure 1c). Data from a total of 123 AMT stations were collected in this study. AMT data observations were recorded at all stations using MTU-5C instruments from Phoenix-Geophysics, Toronto, Canada. At least 1 h of time series data were collected at each station, which were then converted into transfer functions for period ranges from 0.0001 to 10 s.

The data were processed using EMPower software (v2.21.0). The remote referencing technique [36] and Robust estimation [37] were applied at each station. The power spectral density was meticulously analyzed, and data from interference time intervals were eliminated. Despite many mines and considerable human interference in the study area, these technical measures enhanced data processing quality (as shown in Figure 3).

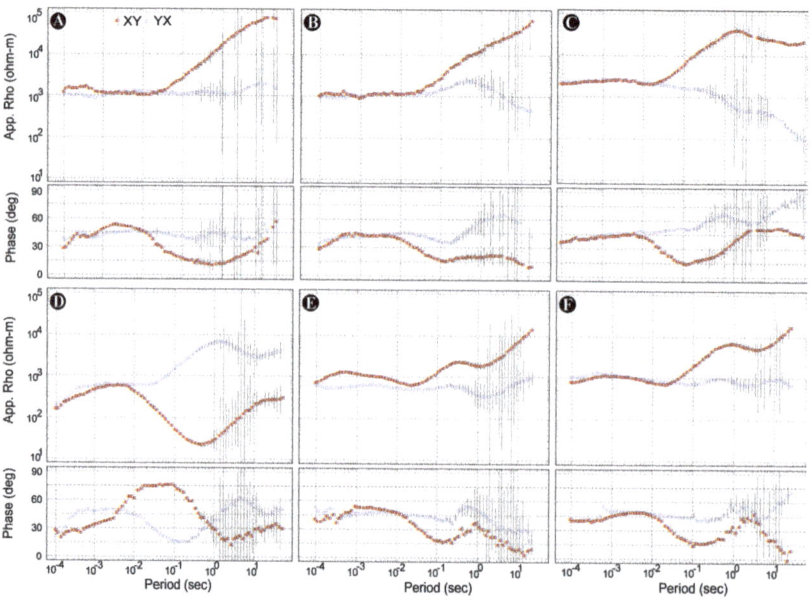

Figure 3. Typical sounding curves at six AMT stations and their locations marked (**A–F**) by the blue dots in Figure 1c.

3.2. Data Analysis

Before inverting the data, dimensionality analysis was conducted to accurately identify the subsurface structures' dimensionality. The skew parameter proposed by Bahr [38] can be used to determine whether EM data satisfy the two-dimensional assumption, which is less affected by the distortion effect of the electromagnetic field. The EM data can be regarded as two-dimensional when the skew value is less than 0.3. At the same time, skew values greater than 0.3 may indicate there are three-dimensional structures in the subsurface. Figure 4 shows pseudo-sections of the Bahr skews, the shallow part of the four profiles exhibits an obvious two-dimensional structure, and the three-dimensional characteristics appear only in the deep part represented by the period 1 s. The profiles satisfy the two-dimensional assumption and can be investigated by two-dimensional inversion.

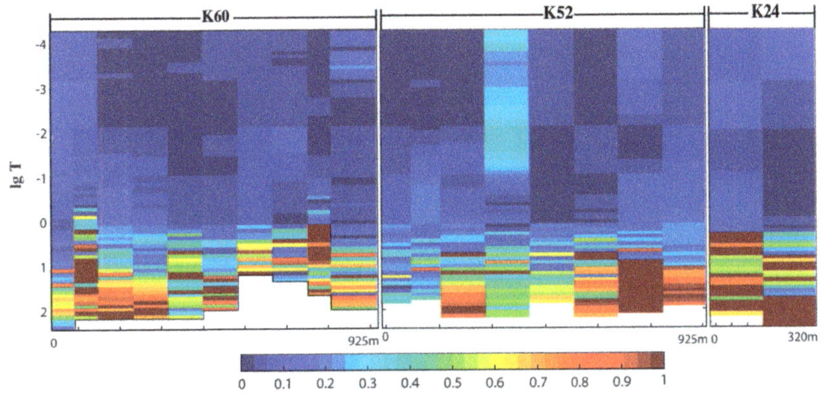

Figure 4. Pseudo-sections of Bahr skew of profiles K60, K52, and K24.

We also used the phase tensor to determine dimensionality. It is the most frequently employed method for dimensionality analysis and has the advantage of being independent of local electric field distortions [39].

The phase tensor is defined by the max phase angle, min phase angle, and skew angle and is shown by the ellipse's major and minor axes and color. If the skew angle is less than 3°, the subsurface medium can be regarded as a two-dimensional structure [40]. In general, the ellipse's major or minor axis represents the direction of the structure, and when the ellipse becomes a circle, the subsurface medium is considered to be one-dimensional.

Figure 5 shows the results of phase tensor analysis for all periods of the three profiles. The skew angles are mostly less than 3°, with some even less than 1°, having equal major and minor axis lengths, suggesting a two-dimensional structure in the shallow portion, with some stations showing one-dimensional characteristics. Only in the last periods (1 s) do many stations have skew angles more significant than 3°. This is consistent with the results of Bahr skews. In summary, the dimensionality of the study area is characterized mainly by 2D features, and a 2D inversion study should be carried out.

After confirming that the two-dimensional inversion is necessary through dimensionality analysis, it is essential to analyze the tectonic strike of the region, and only by rotating the EM data to the direction of the tectonic strike, can the EM data be decomposed into two groups of independent TE and TM polarization modes. The impedance tensor decomposition is widely used to determine the strike direction and remove galvanic distortion in EM data [41]. The rose diagrams in Figure 6 show the strike directions produced by the "strike" algorithm [42]. The strike angle calculated from tensor decomposition may have a 90° ambiguity, which can be solved with induction vectors and regional geological information. Thus, the blue and red wedges perpendicular to each other show two possible strike directions. Considering the results of the surface geological survey, the blue wedges are more consistent with the strike directions. A clearly defined angle of N15°E can be determined for all profiles, which is also consistent with the strike of Au ore belts. Finally, all EM data in profiles k60, k52, and k24 were rotated to the NE15° coordinate system for 2D inversion.

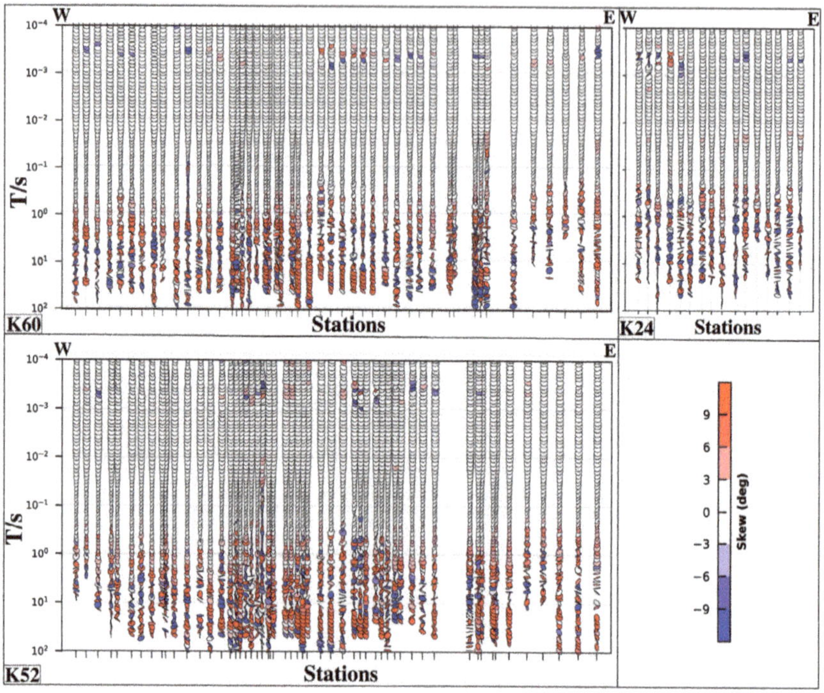

Figure 5. Phase tensor distributions of K60, K52, and K24 AMT profiles.

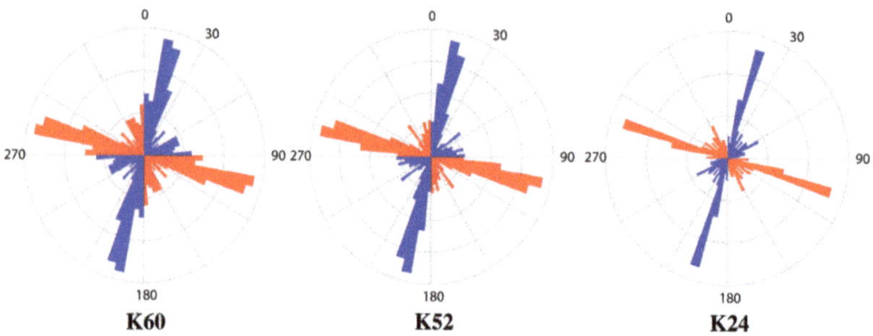

Figure 6. Results of impedance tensor decomposition with data from all periods in each profile.

3.3. 2D Inversions

The 2D inversion uses the idea of multiple modes and multiple parameters combined for multiple 2D inversion trials. To include more data in the inversion and constrain the electrical model, joint inversions of TE + TM were performed using the nonlinear conjugate gradient (NLCG) inversion algorithm [43] integrated into the WinGLink software (2.20.01). AMT data in the period range of 0.0001–1 s were used in the inversion. The initial model was a uniform half-space with a resistivity of 100 Ωm. due to local conductive bodies that shift the apparent resistivity curves' frequency independently but leave the phase curves unaffected [44]. The phase error floors for the TM and TE modes were set to 5%, and the TM resistivity error floor was set at 10%. The TE mode data are easily influenced by the "static shift" effect and the data are difficult to fit in inversion. Therefore, the resistivity error floor for the TE mode was set to 100% to reduce the impact of the static shift. Different regularization factor τ values are used for inversion. After 200 iterations, the final RMS misfit of profiles K60, K52, and K24 are 1.6147, 1.5042, and 1.4231, respectively.

4. Discussion and Interpretation

4.1. Electrical Structural Characteristics

The Mailong mining area has a complex geological structure, with fractured and heavily altered rock formations. There are localized areas of electrical heterogeneity near the shallow surface, providing the basis for this electromagnetic exploration to detect differences in physical properties. By applying precise static shift correction and 2D inversion with terrain data, 2D electrical models for the k60, k50, and k24 AMT profiles in the Miaolong mining area were obtained. The 2D resistivity model aligns well with the distribution of altered zones, rock formations, and faults identified in surface surveys, providing a clear and reasonable depiction of the main geological features and structures in the Miaolong mining area. Comparative studies between the inversion models and known geological structures show that the model can reasonably reflect the actual geological layers and structures.

The K60 profile corresponds to the No.60 exploration line of the Mailong mining area, and the final 2D electrical interpretation model is shown in Figure 7a. The model exhibits a "lateral block, vertical layering" characteristic. Vertically, it can be roughly divided into three layers: a near-surface low-resistivity layer of 5–50 m, a high-resistance layer of 50–800 m, and a deeper low-resistivity layer. The resistivity shows a trend from low to high to medium-low from shallow to deep. The near-surface of the Mailong mining area consists of Quaternary alluvial and residual slope deposits with developed soil layers, abundant vegetation root systems, and relatively thin layer thickness with low resistivity, indicating that the first electrical layer corresponds to the Quaternary cover layer. The second electrical layer is generally thicker, with higher resistivity values typically above 2000 Ωm, comprising mainly of medium-acidic intrusive rocks from the Galidong and Hualixi periods, dominated by granite and diorite, and exhibiting high resistivity. At a depth of around 1000 m, it shows low resistance characteristics. The electrical model is mainly characterized by high-resistivity blocks being intersected by gradient zones of medium to low resistivity. Based on geological data, these shallow electrical gradient zones correspond to known fault structures. The ore-bearing alteration zones in the mining area are mainly strip-like, vein-like, and veinlet-like, mainly occurring in the northeast direction in diorite granite, and are relatively dense with clustered distributions. The related gold ore bodies are all distributed within the alteration zones. Therefore, the low-resistivity and relatively low-resistivity electrical gradient separation zones in the profile reflect the regional structural discontinuity zones formed during early tectonic movements, which are the main controlling structures for ore deposits in the area. This profile runs from west to east, showing multiple low-resistivity zones at 200–220 m, 290–310 m, 400–430 m, and 520–650 m. These low-resistivity anomalies correspond to the four major ore-bearing structural alteration zones (Au VI, Au XV, Au XVI, and Au VIII), mainly discovered through surface surveys and engineering disclosures. However, the low resistivity anomalies corresponding to the ore-bearing structural alteration zones Au VI and Au XV are quite small in scale, with a depth extension of only 20–30 m. Combined with exploration data, the surrounding rock structures in the study area are composed of intermediate-acidic magmatic rocks from the Hualixi period and Precambrian metamorphic rock enclaves. These exhibit primarily high resistivity in terms of electrical properties. At the same time, the mineralized bodies, mainly composed of sulfide-bearing quartz veins filling structural fractures, primarily display low resistance in terms of electrical properties. Considering the geological background of the mining area, rock, and ore electrical properties, it is inferred that the Au VI and Au XV ore zones have a small-scale deep level, with only small low-resistivity anomalies present at the surface. This corresponds with results revealed by surface trench exploration. The Au XVI zone is next in importance but extends only about 200 m. Au VIII has the largest scale, displaying clear low-resistivity anomalies in the deep levels (as shown in Figure 7a(C1)), indicating significant exploration potential and space. Consequently, four boreholes were deployed to verify the deep-level extension situation of the Au VIII zone, and the final verification results were consistent with the model's inference.

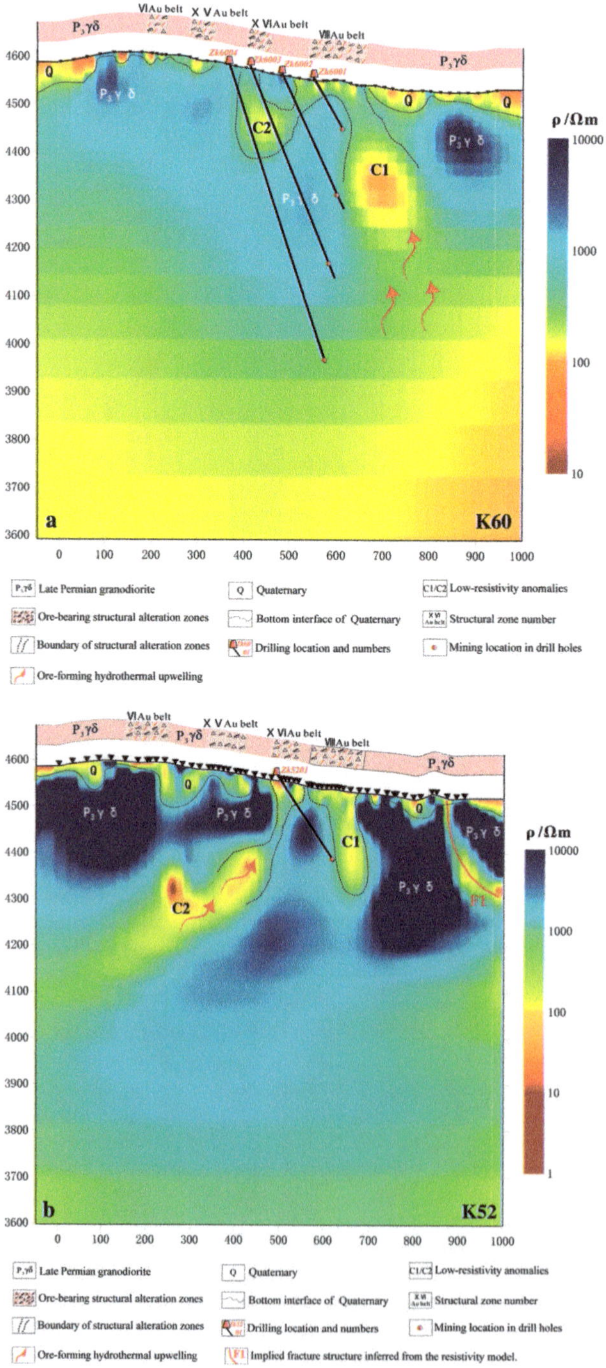

Figure 7. *Cont.*

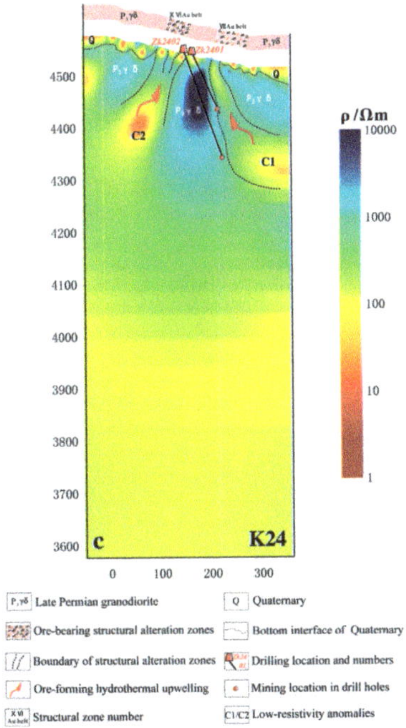

Figure 7. (**a**) Comprehensive geological and geophysical interpretation models of K60. (**b**) Comprehensive geological and geophysical interpretation models of K52. (**c**) Comprehensive geological and geophysical interpretation models of K24.

The K52 profile corresponds to the No.52 exploration line of the Mailong mining area, and the final 2D electrical interpretation model is shown in Figure 7b. This model also exhibits a "lateral block, vertical layering" feature. Vertically, it can be roughly divided into a shallow low-resistivity layer of 5–50 m, a high-resistivity body from 50–900 m with resistivity reaching 10,000 Ωm, and below 900 m, a gradual transition to low-resistivity characteristics, showing a three-layer electrical feature of low–high–low to medium resistivity. The shallow low-resistivity anomaly is inferred to be Quaternary alluvial and residual slope sedimentary cover in the near-surface; the second electrical layer has an overall large thickness with high resistivity values, usually above 4000 Ωm. According to drilling results, this high-resistivity body consists of granites and diorites intruded during the Caledonian and Hercynian periods, mainly showing high resistivity characteristics. The electrical model mainly shows high-resistivity blocks distinctly cut by medium to low-resistivity gradient bands. From west to east in this profile, multiple low-resistivity bands exist at 180–220 m, 350–400 m, 480–520 m, and 600–650 m. These four low-resistivity anomalies correspond to the main ore-bearing structural alteration zones of AuVI, Au XV, AuXVI, and AuVIII, as found in surface surveys and engineering exposures. The ore-bearing alteration zones in this mining area mainly occur in band-like, vein-like, and thin-vein-like forms in the northeastern direction in diorite granite, appearing densely and in clusters within the alteration zones. The related gold ore bodies are all distributed within these alteration zones. Therefore, this profile's low-resistivity and relatively low-resistivity gradient bands reflect the internal structural fault zones and are distinct characteristics of "terminal" ore-bearing structures, with varying depths of different structural fractured alteration zones. In Line 52, the deep extension scale of the AuXVI ore zone is significantly larger than in Line 60, showing clear low-resistivity anomalies in deep parts connected to near-surface low-resistivity trends, hinting at a channel for ore-bearing hydrothermal upwelling based on the ore-forming mechanism of structural altered rock-type gold deposits (shown in

Figure 7b(C2)). On the other hand, the AuVIII ore zone in Line 52 is notably smaller than in Line 60 (shown in Figure 7b(C1)), with surface alteration of the AuVIII ore zone observed in early drill hole 5201 data but not extensively extending into deeper parts, aligning with the model's speculation in this study. The inconsistency in the scale of the AuVIII ore zone over such a short distance indicates complex tectonic movements in the study area, leading to varied fault fracturing, multiple-stage alteration, and erosion, providing different mineralization spaces for hydrothermal activities, thereby suggesting that the ore bodies in the study area might have segmented enrichment characteristics. Additionally, new hidden structures were discovered in the profile, likely reflecting fractured alteration zones, yet due to the coverage of the fourth series at the surface, this area remains a target zone for further exploration.

The K24 profile corresponds to exploration line No.24 in the Mailong mining area, and the final 2D electrical interpretation model is shown in Figure 7c. This model can be roughly divided vertically into a shallow low-resistivity layer of 5–20 m, a high-resistivity body reaching 10,000 Ωm from 20–500 m, and a gradually appearing low-resistivity feature below 500 m, showing a three-layer electrical feature of low–high–medium resistivity. The shallow low-resistivity anomaly is speculated to be the cover layer of the fourth series; the second electrical layer has a generally thick thickness, with high resistivity values typically above 3000 Ωm, reflecting the medium-acid intrusive rock mass of the Hualixi period, mainly showing high resistivity features. Laterally, the electrical structure mainly shows segmented features of high resistivity and medium-low resistivity anomalies. Based on previous research and analysis, it is found that the high resistivity–low resistivity electrical gradient zone is a fault fracture zone formed by structural movement inside the area. The profile intersects the known AuXVI and AuVIII ore zones in the area, and the ore zone positions coincide with the gradient zones in the section. The low-resistivity zone is considered to be related to the distribution of structures and ore zones, especially the significant low-resistivity anomalies deep in the AuXVI and AuVIII ore zones (shown in Figure 7c(C1,C2)) connected to the shallow low-resistivity trend, possibly indicating a structural feature of the fracture zone extending deep underground and providing a channel for hydrothermal uplift for epithermal gold deposits based on structural alteration.

4.2. Deep Prospecting Breakthrough

From the 2D resistivity model, we found that the location of the ore body is the high and low resistance contact conversion belt. The depth of each ore zone is roughly the same, and all of them tend to extend to the depth. The low-resistivity bands respond to the region's major tectonic fracture alteration zones. Especially in the K60 profile, there is a clear low-resistivity zone deep in the AuVIII ore belt, which is speculated to reflect the deep structural fractured zone. There may be concealed mineralized bodies in the deep, indicating a good prospecting outlook in the deep part of this area and delineating favorable areas for deep exploration. Subsequently, drill hole 6004 confirmed the presence of a 1.67 m ore body at a depth of 660 m, with a gold grade of 2.37 g/t, effectively validating the geophysical and geochemical exploration methods. It is believed that the deep part of this area has good potential for mineral exploration, and it is speculated that deep borehole verification of over 1000 m can be carried out. Similarly, in the K52 profile, a significant low-resistivity anomaly was discovered deep in the AuXVI ore belt (as shown in Figure 7b(C1)), which can serve as a deep exploration target area. Due to this, priority will be given to verifying this area in later work. Additionally, in K24, the thick Quaternary cover has hindered surface operations from progressing. However, a 2D electrical profile passing through the covered area effectively inferred a low-resistivity fault zone. It was found that the AuXVI and AuVIII mineral belts extend southward to the vicinity of the 24th exploration line. Through surface trenching, it was revealed to be a structurally altered zone with apparent gold mineralization characteristics, serving as a basis for further verification of the deep target area (as shown in Figure 7c(C1)). As a result, the electromagnetic method has yielded significant results in exploring the Golou gold deposit, marking a breakthrough in the exploration work.

4.3. Mineralization Mechanism and Model

After years of geological exploration and scientific research, previous researchers extensively analyzed and studied the magmatic activity and gold mineralization in the Gouli area from the perspectives of the mineralization era, mineralizing fluids, and mineral sources. It is widely believed that the formation of gold deposits (ore bodies) in the Gouli area is closely related to regional magmatic tectonic activity and exhibits characteristics of multiple mineralization phases [18,21,23,33]. This study suggests that the intense tectonic movements during the Caledonian to Hercynian period in the Golol region led to frequent and vigorous magma activities, forming the Kunbei magmatic arc zone. The Kunbei magmatic arc zone, formed by tectonic magmatic activities, was influenced by extension and compression from the Central Tethys Ocean during the Indonesian period. This resulted in the formation of multiple sets of fracture structures in different directions and with different mechanical properties due to north–south compressional stress in the region. Additionally, the continued thickening of the Kunbei magmatic arc zone triggered the crust and mantle remelting, resulting in deep-seated molten magma formation. This molten magma intruded along the fracture structures, promoting the intrusion and migration of gold-rich magmatic–hydrothermal fluids along these fracture structures. In regions where there were abrupt changes in physical and chemical conditions along major ductile fracture structures, mineral-rich hydrothermal fluids boiled, causing reactions with surrounding rocks and rapid deposition and enrichment of ore-forming materials like gold, ultimately forming corresponding gold ore bodies (as shown in Figure 8).

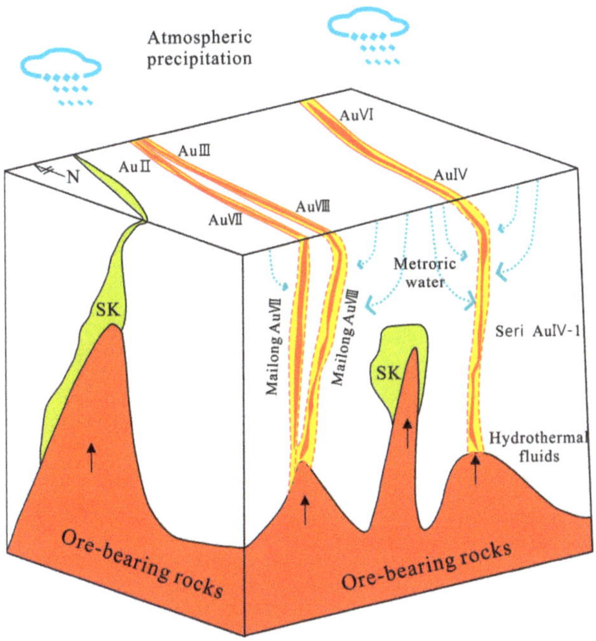

Figure 8. Mineralization mechanism and model of the Mailong mining area.

5. Conclusions

This paper constructed a 2D resistivity model of the Mailong mining area based on audio-frequency magnetotelluric data in the Gouli ore-concentration area in the eastern Kunlun metallogenic belt. Based on the results of the surface geological survey, drill hole verification, and other data, the deep extension of the main ore belt was discussed, and the deep exploration target area was delineated. A breakthrough in deep mineral exploration was achieved through later drill hole verification. Finally, the electrical constraints, such as mineralization regularities and material sources, were provided for the mineralization mechanism of the Mailong mining area. The main conclusions are drawn as follows:

(1) The study area is controlled by three major fracture structures of the Northern Kunlun Fault, the Central Kunlun Fault, and the Southern Kunlun Fault, with frequent and intense magmatic activities mainly during the Indosinian–Yanshanian period. This has resulted in the fractured basement, extensive outcrops of intrusive rocks, and a network of mainly NNW–NE strike-slip and thrust faults crisscrossing the area, creating excellent ore-forming geological conditions within the region.

(2) The 2D resistivity model reveals that different ore belts in the area extend to varying depths at the same location, and the same ore belts have different extents at different locations. This indicates that the study area has undergone complex tectonic movements resulting in varying fault displacements and multiple phases of alteration and erosion, providing different ore-controlling spaces for mineralization hydrothermal activities, ultimately resulting in segmented enrichment characteristics of ore bodies in the study area.

(3) This electromagnetic exploration defined a favorable deep mineralization zone. Subsequent drilling verification revealed a gold-bearing alteration structure. Therefore, electromagnetic exploration achieved good results in exploring structurally altered rock-type gold deposits in the plateau desert area, leading to a breakthrough in deep mineral exploration.

Author Contributions: J.D.: Conceptualization, Resources, Writing—review & editing, Funding acquisition, Project administration, Investigation. P.W.: Conceptualization, Methodology, Data curation, Writing—original draft. H.L.: Supervision, Resources, Writing—review & editing. H.Z.: Writing—review & editing. M.Z.: Writing—review & editing, Investigation. H.T.: Writing—review & editing, Funding acquisition, Project administration. X.Y.: Writing—review & editing. J.L.: Writing—review & editing. B.Z.: Writing—review & editing. All authors have read and agreed to the published version of the manuscript.

Funding: This research is supported by Qinghai Provincial Association for Science and Technology Youth Science and Technology Talent Support Project (2023QHSKXRCTJ47), Qinghai Province "Kunlun Talent High Innovation and Entrepreneurship Talent" Project and Demonstration of Deep Exploration Breakthrough in Important Mineral Concentration Areas of Qinghai Province (2023085029ky004), Exploration Foundation of Qinghai Province (2023085029ky004).

Data Availability Statement: Data will be made available on request.

Acknowledgments: We thank Gary Egbert and Alan Jones for ModEM and MTmap.

Conflicts of Interest: The authors declare no conflicts of interest.

References

1. Li, Z.; Xue, C.; Wang, X.; Tang, H.; Tu, Q.; Teng, J.; Li, R. Features of Reqional Mineralization and Analysis of the Exploration Development in the Eastern Kunlun Mountains. *Geol. Rev.* **2007**, *53*, 708–717.
2. Guo, X.; Jia, Q.; Li, J.; Kong, H.; Yao, X.; Mi, J.; Qian, B.; Wang, Y. Zircon U-Pb geochronology and geochemistry and their geological significances of eclogites from East Kunlun high-pressure metamorphic belt. *Earth Sci.* **2018**, *43*, 4300–4318.
3. Xiong, W.; Zhao, M.; Yu, M. Ocean–Continent Transition Process and Magmatism in Orogenic Belts: A Case Study of Paleozoic Granites in the Dulan Area of East Kunlun. *Northwest. Geol.* **2023**, *56*, 113–139.
4. Du, Y.-L.; Jia, Q.-Z.; Han, S.-F. Mesozoic Tectono-Magmatic-Mineralization and CopperiGold Polymetallic Ore Prospecting Research in East Kunlun Metallogenic Belt in Qinghai. *Northwest. Geol.* **2012**, *45*, 69–75.
5. Xu, Z.; Yang, J.; Li, H.; Zhang, J.; Zeng, L.; Jiang, M. The Qinghai G Tibet Plateau and Continental Dynamics: A Review on Terrain Tectonics, Collisional Orogenesis, and Processes and Mechanisms for the Rise of the Plateau. *Geol. China* **2006**, *33*, 221–238.
6. Luo, Z.; Deng, J.; Cao, Y.; Guo, Z.; Mo, X. On late Paleozoic-early Mesozoic volcanism and regional tectonic evolution of Eastern Kunlun, Qinghai Province. *Geoscience* **1999**, *13*, 51–56.

7. Mo, X.; Luo, Z.; Deng, J.; Yu, X.-H.; Liu, C.; Yuan, W.; Bi, X. Granitoids and crustal growth in the East-Kunlun orogenic belt. *AGU Fall Meet. Abstr.* **2011**, *2011*, T51D-2370.
8. Dong, Y.; He, D.; Sun, S.; Liu, X.; Zhou, X.; Zhang, F.; Yang, Z.; Cheng, B.; Zhao, G.; Li, J. Subduction and accretionary tectonics of the East Kunlun orogen, western segment of the Central China Orogenic System. *Earth Sci. Rev.* **2018**, *186*, 231–261. [CrossRef]
9. Liu, Y.; Chen, Z.; Li, W.; Xu, X.; Kou, X.; Jia, Q.; Zhang, Z.; Liu, F.; Wang, Y.; You, M. The Cu-Ni mineralization potential of the Kaimuqi mafic-ultramafic complex and the indicators for the magmatic Cu-Ni sulfide deposit exploration in the East Kunlun Orogenic Belt, Northern Qinghai-Tibet Plateau, China. *J. Geochem. Explor.* **2019**, *198*, 41–53. [CrossRef]
10. Zhang, J.; Lei, H.; Ma, C.; Li, J.; Pan, Y. Geochemical and thermodynamical modeling of magmatic sources and processes for the Xiarihamu sulfide deposit in the eastern Kunlun Orogen, western China. *J. Geochem. Explor.* **2018**, *190*, 345–356. [CrossRef]
11. Chen, J.; Fu, L.; Wei, J.; Tian, N.; Xiong, L.; Zhao, Y.; Zhang, Y.; Qi, Y. Geochemical characteristics of late ordovician granodiorite in Gouli Area, eastern Kunlun orogenic belt, Qinghai Province: Implications on the evolution of Proto-Tethys ocean. *Earth Sci.* **2016**, *41*, 1863–1882.
12. Liang, L.; Fengyue, S.; Shijin, L.; Bile, L.; Ye, Q.; Chao, W.; Tuofei, Z.; Lu, Y.; Guan, W.; Liang, H.; et al. Metallogenic Geological Conditions and Regularity of Magmatic Cu -Ni SulfideDeposits in the East Kunlun Metallogenic Belt. *J. Jilin Univ. Earth Sci. Ed.* **2022**, *52*, 1461–1496.
13. Norbu, N.; Tang, L.; Li, J.; Kong, H.; Li, Y.; Jia, Q.; Xu, Y. Petrogenesis of Middle Triassic Adakite-like Intrusions in the Asiha Orogenic Gold Deposit, East Kunlun Orogenic Belt, China. *Minerals* **2023**, *13*, 74. [CrossRef]
14. Feng, C.; Zhang, D.; Wang, F.; She, H.-Q.; Li, D.-X. Multiple orogenic processes and geological characteristics of the major orogenic gold deposits in East Kunlun area, Qinghai Province. *Acta Geosci. Sin.* **2004**, *25*, 415–422.
15. He, S.; Lin, G.; Zhong, S.; Bai, G.; Zhang, Y.; Shu, S.; Wang, L.; Furong, Z. Geological Characteristics and Related Mineralization of "Qinghai Gold Belt" Formed by Orogeny. *Northwest. Geol.* **2023**, *56*, 1–16.
16. Li, J.; Du, W.; Cheng, Y.; Kong, H.; Liu, J.; Li, Y.; Jia, Q.; Namkha, N.; Xia, M.; Li, Y. Characteristics of gold deposits and ore-control factors in the East Kunlun mineralization belt, Qinghai Province. *Geol. Explor.* **2015**, *51*, 1079–1088.
17. Wanming, Y.; Xuanxue, M.; Xuehui, Y.; Zhao-Hua, L.; Wei-Dong, G. Gold metallogenic belts and prospecting direction in eastern Kunlun area. *Geol. Explor.* **2000**, *36*, 20–23.
18. Yang, B.; Zhang, L.; Ma, Z.; Wang, X. Study on Metallogenic Geological Background of Gold Deposits in Gouli Area. *Qinghai. Min. Explor.* **2018**, *9*, 1920–1925.
19. Yuan, W.; Mo, X.; Wang, S.; Zhang, X. The relationship between gold mineralization and regional tectonic evolution in eastern kunlun mountains. *Geol. Prospect.* **2003**, *39*, 5–8.
20. Li, S.; Zeng, X.; Wang, F.; Tong, H.; Yang, Y.; Zhang, D.; Kang, B.; Shi, H.; Zheng, Z. Main metallogenic series and prospecting potential in Qinghai province. *J. Jilin Univ. Earth Sci. Ed.* **2022**, *52*, 1419–1445.
21. Zhang, B.; Ma, Q.; Zhang, L. Cause analysis of geological characteristics of gold and polymetallic ore deposit in Qinghai county of Dulan province Mairon area. *World Nonferrous Met.* **2018**, *7*, 131–132.
22. Dong, M.; Guosheng, S.; Xue, L.; Lei, H.; Jitian, Z.; Geny, L. Geological characteristics and genesis of gold deposits in the Gouli area, Dulan County, Qinghai Province. *GOLD* **2022**, *11*, 13–18.
23. Yue, W.; Zhou, J.; Gao, J.; Huang, Y.; Jia, F. Geochemistry, Zircon U-Pb Dating and Hf Isotopic Compositions of the Granite-porphyry in Asiha Gold Ore District, Dulan County, Qinghai Province. *Geotecton. Metallog.* **2017**, *41*, 776–789.
24. Song, S.; Niu, Y.; Su, L.; Xia, X. Tectonics of the North Qilian orogen, NW China. *Gondwana Res.* **2013**, *23*, 1378–1401. [CrossRef]
25. Wei, W. New Advance and Prospect of Magnetotelluric Sounding (MT) in China. *Prog. Geophys.* **2002**, *17*, 245–254.
26. Wei, W.-B.; Jin, S.; Ye, G.-F. Methods to study electrical conductivity of continental lithosphere. *Earth Sci. Front.* **2003**, *10*, 15–24.
27. Tezkan, B.; Hordt, A.; Gobashy, M. Two-dimensional radiomagnetotelluric investigation of industrial and domestic waste sites in Germany. *J. Appl. Geophys.* **2000**, *44*, 237–256. [CrossRef]
28. Li, F.; Zeng, Q.; Zhu, R.; Chu, S.; Xie, W.; Zhang, B.; Zhang, X. Application of the AMT Method to Gold Deposits: A Case Study in the Qinling Metallogenic Belt of North China Craton. *Minerals* **2021**, *11*, 1200. [CrossRef]
29. Pitiya, R.; Lu, M.; Chen, R.; Nong, G.; Chen, S.; Yao, H.; Shen, R.; Jiang, E. Audio Magnetotellurics Study of the Geoelectric Structure across the Zhugongtang Giant Lead–Zinc Deposit, NW Guizhou Province, China. *Minerals* **2022**, *12*, 1552. [CrossRef]
30. Qi, P.; Yin, Y.; Jin, S.; Wei, W.; Xu, L.; Dong, H.; Huang, J. Three-Dimensional audio-magnetotelluric imaging including surface topography of the cimabanshuo porphyry copper deposit, Tibet. *Minerals* **2021**, *11*, 1424. [CrossRef]
31. Xinluo, Y.; Xiaozhu, L.; Bin, L.; Jiyong, C.; Guicong, F. Application of the audio magnetotelluric (AMT) method to the exploration of the Dingtou gold deposit in Fengshan County, Guangxi. *Geol. Explor.* **2021**, *57*, 1366–1373.
32. Zhuoyi, W.; Jiuda, S.; Zhiqiang, Z.; Guosheng, S.; Xue, L.; Shengbao, W.; Dong, M. Geological and geochemical anomaly characteristics and geological significance of Seri Gold-Silver Deposit in Dulan County, Qinghai Province. *GOLD* **2023**, *2*, 73–77.
33. Zhiyuan, S.; Tongke, W.; Guozheng, J.; Xinming, Z.; Xu, W. Analysis of ore prospecting of Mailong-Seri area, Dulan county Qinghai province and the target delineation. *Contrib. Geol. Miner. Resour. Res.* **2022**, *37*, 291–301.
34. He, J.; Yang, S.; Duan, H.; Liu, Y.; Xiong, S. Qinghai Serir Gold Mine Ore-Forming Characteristics and Prospects for Mineral Exploration. *Min. Technol.* **2022**, *22*, 238–243.
35. Chave, A.D.; Jones, A.G. *The Magnetotelluric Method: Theory and Practice*; Cambridge University Press: Cambridge, UK, 2012.
36. Gamble, T.D.; Goubau, W.M.; Clarke, J. Magnetotellurics with a remote magnetic reference. *Geophysics* **1979**, *44*, 53–68. [CrossRef]

37. Egbert, G.D.; Booker, J.R. Robust estimation of geomagnetic transfer functions. *Geophys. J. R. Astron. Soc.* **1986**, *87*, 173–194. [CrossRef]
38. Bahr, K. Geological noise in magnetotelluric data: A classification of distortion types. *Phys. Earth Planet. Inter.* **1991**, *66*, 24–38. [CrossRef]
39. Caldwell, T.G.; Bibby, H.M.; Brown, C. The magnetotelluric phase tensor. *Geophys. J. Int.* **2004**, *158*, 457–469. [CrossRef]
40. Booker, J.R. The magnetotelluric phase tensor: A critical review. *Surv. Geophys.* **2014**, *35*, 7–40. [CrossRef]
41. Groom, R.W.; Bailey, R.C. Decomposition of magnetotelluric impedance tensors in the presence of local three-dimensional galvanic distortion. *J. Geophys. Res.* **1989**, *94*, 1913–1925. [CrossRef]
42. McNeice, G.W.; Jones, A.G. Multisite, multifrequency tensor decomposition of magnetotelluric data. *Geophysics* **2001**, *66*, 158–173. [CrossRef]
43. Rodi, W.; Mackie, R.L. Nonlinear conjugate gradients algorithm for 2-D magnetotelluric inversion. *Geophysics* **2001**, *66*, 174–187. [CrossRef]
44. Jones, A.G. Electromagnetic interrogation of the anisotropic Earth: Looking into the Earth with polarized spectacles. *Phys. Earth Planet. Inter.* **2006**, *158*, 281–291. [CrossRef]

minerals

MDPI

Article

Moment Estimation from Time Domain Electromagnetic Data

Marc A. Vallée * and Mouhamed Moussaoui

Geo Data Solutions GDS Inc., 4402 Louis-B. Mayer, Laval, QC H7P 0G1, Canada
* Correspondence: mvallee@geodatasolutions.ca

Abstract: Moment representations have been proposed to facilitate the interpretation of geophysical time domain electromagnetic responses. We present a new methodology for estimating these moments from field data for different system waveforms when on-time and off-time measurements are available. Quadrature impulse response moments are estimated by a recursive relation involving moments of the input waveform and moments of the observed response. After adapting this method to time domain electromagnetic applications—in particular, MEGATEM and AeroTEM (AirTEM) airborne electromagnetic systems—we present the results from applying this method on synthetic and real data collected over the Reid–Mahaffy test site in northern Ontario, Canada.

Keywords: electromagnetic exploration; time domain measurements; moments

1. Introduction

Time domain electromagnetic (TDEM) methods are used in exploration geophysics to search for subsurface mineralized zones that are conductive [1]. Their principle is as follows: an electric current pulse, associated with a primary magnetic field through Ampère's law, circulates in the transmitter loop, as illustrated in Figure 1. According to Faraday's law of induction, this primary magnetic field generates secondary currents in conductive materials in the subsurface. These secondary currents are associated with a secondary magnetic field that is detected by the receiver loop. As the pulse is of limited duration, the secondary signal can be measured while the pulse is on (on-time) and after the pulse is turned off (off-time). Typically, only off-time measurements are recorded since on-time measurements are affected by the primary field. In airborne electromagnetics (AEMs), the transmitter (source) and receiver are either located on the aircraft or are towed below the aircraft (Figure 2).

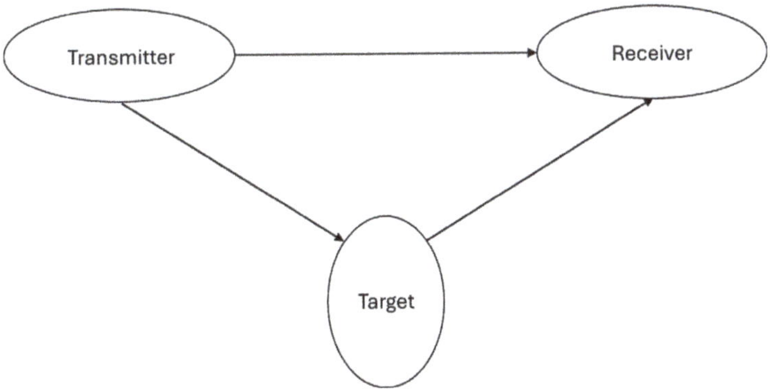

Figure 1. Schematic of a transmitter–receiver target system.

Figure 2. Example of a helicopter TDEM system (AirTEM) towed below the aircraft. The receiver is located on the same bird as the transmitter.

There have been many advancements since the early development of TDEM techniques to localize mineralized zones with EM measurements. Better understanding of the electromagnetic (EM) principles involved [1–3] and the development of computer techniques have led to improvements in various domains. Ref. [4] proposed to localize anomalies and interpret them by comparing them with the results of analog model simulations. Following advances in computer technology, Ref. [5] developed an inversion program to invert AEM data or to retrieve the parameters of the model by numerical curve-fitting using the numerical model of a plate in free space developed by [6]. The model based on the inversion of a plate was extended by [7] to consider host rock to be conductive, layered earth. Refs. [8,9] proposed an algorithm for inverting a sphere in layered earth. The thin sheets in the layered earth model of [7] were included by [10] for an initial model for the probabilistic inversion of AEM data for basement conductors.

Some of these elements were incorporated into the three-step process proposed by [11] and refined by [12], which is as follows: (1) compress the data to decay time-constant space, (2) transform the response to a conductivity–depth image, and (3) parameterize local anomalies with plate-like conductors. Similarly, Ref. [13] proposes to invert AEM data in two phases in order to determine background conductivity models with layered-earth inversions and interpret discrete conductors as electric or magnetic dipoles, while the approach in [14] approximates the subsurface current system using a 3D subsurface grid of either 3D magnetic or electric dipoles, for which locations and orientations are solved by inversion. Some of the most elaborate interpretation techniques invert the complete airborne dataset with sophisticated mathematical techniques and solve the forward model with an integral equation [15–17], finite volume method [18], or randomly undersampled forward model [19].

Despite these advances, elaborate EM interpretation techniques are not generalized to all AEM datasets as they are complex and costly in terms of time and computer resources and because of the enormous dataset sizes. For these reasons, simple interpretation tools have the potential of assessing data quality quickly, locating discrete conductors, and facilitating the use of more complex interpretation methods. Some of these tools include the use of parameters such as the apparent conductivity of the subsurface [20,21], the decay time constant [22,23], and inductive and resistive limits [24]. At the inductive limit, current flows do not penetrate conductive bodies, and at the resistive limit, the response varies linearly with conductivity. Resistive limit estimation was used by [21,24,25] as the first step of an EM interpretation process. Refs. [26,27] extended the concepts of these limits to the moments of the quadrature impulse response, which are time integrals of the response signal, as the zero-order and first-order moments correspond to the inductive and resistive limit, respectively. Subsequent authors developed solutions for the moments of a sphere in free-space [27], in homogeneous half-space and excited by a source located on the surface of the ground [28,29], thick horizontal layer [30], and in one-dimensional earth with finite conductance [31]. In the year 2000, CGG and Geophysical Software Solutions collaborated to develop an application called EM-Q based on the research done by [32] on the uniform field approximation of the response of a sphere, for which [8] compared the results to the solution attained in [5]. The moments were also used by [25] as an approximation for the development of a 3D EM inversion program.

However, as TDEM measurements are done with systems using different primary current waveforms, the estimation of the moments from the quadrature impulse response is sometimes problematic. Ref. [27] developed a solution by approximating the TDEM waveform as a simple current box and applied this technique to the data integral from a MEGATEM AEM system [33], for which the current waveform is a half-sine. However, this approach is not applicable to more elaborate waveforms such as those of the AeroTEM AEM system [34], for which the current waveform is a symmetrical triangular pulse. Furthermore, Ref. [35] encountered difficulties with the definition of zero time while analyzing MEGATEM data, which resulted in a noisy moment estimation. Ref. [28] only considers step-response measurements, from which it is possible to estimate moments directly. In conclusion, although there are extensive theoretical developments on the use of moments in the field of TDEM interpretation, estimation of the moments from various waveforms on synthetic or field data has been neglected, slowing the application of this method on real data.

For these reasons, we propose a method for directly estimating the quadrature impulse moments from any waveform and measured signals. This method should increase the use of the moment method and better outline its limitations. The proposed algorithm evaluates the moments of the quadrature impulse response with a recursive relation involving the moments of the pulse current derivative and the moments of the quadrature on-time and off-time responses. Consequently, it requires that on-time measurements are available. After a noise analysis, we apply this method to synthetic AirTEM results (the same waveform as AeroTEM) with a sphere model in layered earth, illustrating the conditions for which moments can be useful for outlining a discrete conductor. This is followed by application of the moment method on field data from MEGATEM and AirTEM systems collected over the Reid–Mahaffy test site, Ontario, Canada.

2. Theory

2.1. Moment Estimation

The quadrature impulse response of a TDEM system can be developed as follows [27]. The EM step response of the subsurface can be written as

$$s(t) = -Bu(t)h(t) \tag{1}$$

where t is time, B is the response at $t = 0$, $u(t)$ is the unit step-on function, and $h(t)$ is a dimensionless function characterizing the secondary decay associated with currents in the ground. The total impulse response $r(t)$ is the time derivative of $s(t)$:

$$r(t) = -B\left\{\delta(t)h(t) + u(t)\frac{\partial h(t)}{\partial t}\right\} \tag{2}$$

Analogous to frequency-domain systems, Ref. [26] names the first and second terms of this equation the in-phase and quadrature components, respectively, and focuses on the second term. This is represented as

$$i(t) = -Bu(t)\frac{\partial h(t)}{\partial t} \tag{3}$$

The moments of the quadrature impulse response $i(t)$ are defined as

$$I^n = \int_0^\infty t^n i(t)dt \tag{4}$$

In practice, the primary waveform is different from a step response and is convolved with the impulse response. Furthermore, the receiver coil detects the derivative of the magnetic field. This can be written as

$$y(t) = x(t) * i(t) \tag{5}$$

where $x(t)$ represents the waveform derivative, $*$ is the convolution operator, and $y(t)$ is the measured secondary field, from which the on-time primary field and the secondary in-phase field have been removed. This equation represents the result $y(t)$ of a causal convolution of the system waveform $x(t)$ and the system transfer function $i(t)$, where $x(t) = i(t) = 0$ for $t < 0$

$$y(t) = \int_0^\infty i(h)x(t-h)dh = \int_0^\infty i(t-h)x(h)dh \tag{6}$$

Following [36,37], we develop the moments of the convolution response as

$$Y^n = \int_0^\infty y(t)t^n dt = \int_0^\infty \int_0^\infty i(h)x(t-h)t^n dhdt \tag{7}$$

If we replace $u = t - h$, then $t = u + h$ and $du = dt$; then, the expression becomes

$$Y^n = \int_{-h}^{\infty-h} \int_0^\infty i(h)x(u)(u+h)^n dhdu \tag{8}$$

Since $x(u) = 0$ for $u < 0$ and $\infty - h \simeq \infty$, we can re-write the equation as

$$Y^n = \int_0^\infty \int_0^\infty i(h)x(u)(u+h)^n dhdu \tag{9}$$

Developing the power of a binomial expansion, it is possible to separate the two integrals in the following way:

$$\begin{aligned} Y^n &= \int_0^\infty \int_0^\infty i(h)x(u)\left[\sum_{k=0}^n \binom{n}{k} u^{n-k}h^k\right] dhdu \\ Y^n &= \sum_{k=0}^n \binom{n}{k} \int_0^\infty x(u)u^{n-k}du \int_0^\infty i(h)h^k dh \end{aligned} \tag{10}$$

where

$$\binom{n}{k} = \frac{n!}{k!(n-k)!} \tag{11}$$

If we define the input and impulse response moments as

$$X^n = \int_0^\infty x(t)t^n dt \tag{12}$$

and

$$I^n = \int_0^\infty i(t)t^n dt, \tag{13}$$

we can write

$$Y^n = \sum_{k=0}^{n} \binom{n}{k} X^{n-k} I^k. \tag{14}$$

For $X^0 \neq 0$, if we develop

$$Y^n = \sum_{k=0}^{n-1} \binom{n}{k} X^{n-k} I^k + X^0 I^n, \tag{15}$$

we can evaluate I^n from X^k and Y^k, $k \in [0, n]$, as

$$I^n = \frac{Y^n - \sum_{k=0}^{n-1} \binom{n}{k} X^{n-k} I^k}{X^0}. \tag{16}$$

However, in the case of some AEM systems, the current waveform or derivative is symmetric, with positive and negative lobes, such as in MEGATEM and AeroTEM, and $X^0 = 0$. Then we need to use the following relation:

$$Y^n = \sum_{k=0}^{n-2} \binom{n}{k} X^{n-k} I^k + \binom{n}{n-1} X^1 I^{n-1} \tag{17}$$

from which we deduce

$$I^{n-1} = \frac{Y^n - \sum_{k=0}^{n-2} \binom{n}{k} X^{n-k} I^k}{\binom{n}{n-1} X^1} \tag{18}$$

or

$$I^n = \frac{Y^{n+1} - \sum_{k=0}^{n-1} \binom{n+1}{k} X^{n+1-k} I^k}{\binom{n+1}{n} X^1} \tag{19}$$

The next subsections demonstrate how this relation can be applied to the MEGATEM and AeroTEM systems.

2.2. MEGATEM

MEGATEM [33] is a fixed-wing system with a transmitter loop wound around the nose, wingtips, and tail of a DeHavilland Dash-7. The transmitter loop covers an area of 406 m^2, while the receiver comprises three magnetic coils that are towed 50 m below and 131 m behind the aircraft. As stated before, the current pulse is a half-sine. The primary field extraction [38], also called stripping, is done using a reference waveform that is measured at altitude and by minimizing the functional Φ, where

$$\begin{aligned}\Phi &= \textstyle\sum_n [T(t_i) - \alpha R(t_i)] \\ &= \textstyle\sum_n Q(t_i)\end{aligned} \tag{20}$$

$T(t_i)$ is the total field response, $R(t_i)$ is the reference waveform, and α is an unknown factor that is estimated from the minimization. As [38] points out, this procedure estimates the on-time in-phase and quadrature responses. Accordingly, this approach provides the information required for moment estimation using the proposed method.

2.3. AeroTEM

The AeroTEM system [34] is a helicopter-towed EM system with coincident transmitter and receivers characterized by a triangular current pulse (Figure 3) for which the derivative is a constant with alternating polarity. The voltage sensed by the receiver, which is surrounded by a bucking coil to cancel the primary field, is a series of step-responses, i.e., a first step of positive polarity, a second step of double-negative polarity, and a last step of positive polarity. The on-time field is measured as a sum of two voltages separated by half the period.

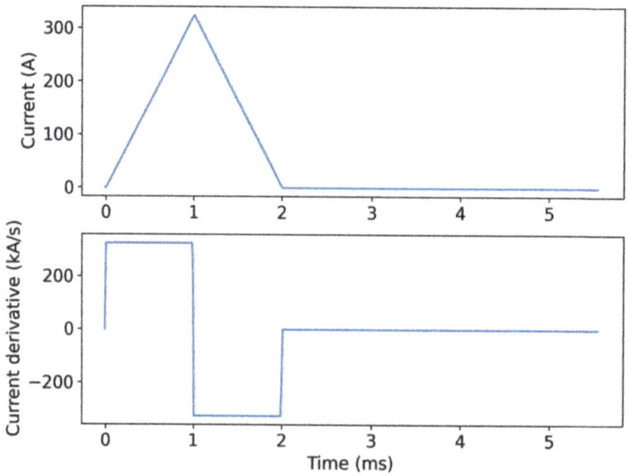

Figure 3. Waveform and derivative waveform of the AeroTEM system.

The response derivative can be seen as the sum of three step-response of alternating polarities and is expressed as

$$z(t) = -Bu(t)h(t) + 2Bu(t - P/2)h(t - P/2) - Bu(t - P)h(t - P) \tag{21}$$

where P is the pulse width, and B, $u(t)$, and $h(t)$ are defined as in Equation (1). This can be decomposed according to time:

$$\begin{aligned} z(t) &= -Bh(t) && 0 < t < P/2 \\ &= -B[h(t) - 2h(t - P/2)] && P/2 < t < P \\ &= -B[h(t) - 2h(t - P/2) + h(t - P)] && P < t \end{aligned} \tag{22}$$

The in-phase $x(t)$ and quadrature $y(t)$ responses are, respectively,

$$\begin{aligned} x(t) &= -Bh(0) && 0 < t < P/2 \\ &= Bh(0) && P/2 < t < P \\ &= 0 && P < t \end{aligned} \tag{23}$$

and

$$\begin{aligned} y(t) &= -B[h(t) - h(0)] && 0 < t < P/2 \\ &= -B[h(t) - 2h(t - P/2) + h(0)] && P/2 < t < P \\ &= -B[h(t) - 2h(t - P/2) + h(t - P)] && P < t \end{aligned} \tag{24}$$

The on-time sum measurement typical of the AeroTEM system is

$$\begin{aligned} d(t) &= z(t) + z(t + P/2) \\ &= -Bh(t) - B[h(t + P/2) - 2h(t)] \\ &= -B[h(t + P/2) - h(t)] \end{aligned} \tag{25}$$

If we assume that $h(t + P/2) << h(t)$, we can approximate that $d(t) \simeq Bh(t)$. Using this approximation, $y(t)$ can be estimated from Equation (23) as

$$\begin{aligned} y(t) &\simeq d(0) - d(t) \\ y(t + P/2) &\simeq 2d(t) - d(0) \end{aligned} \tag{26}$$

2.4. Incomplete Moments

In practice, response moments are estimated between the early and late times, as represented by the following equation [28]:

$$Y^n = \int_0^{t_1} y(t)t^n dt + \int_{t_1}^{t_n} y(t)t^n dt + \int_{t_n}^{\infty} y(t)t^n dt \tag{27}$$

where [28] labels the first and last terms the "head" and the "tail", respectively, and the middle term is the incomplete moment. As demonstrated by [28], if the waveform is a step, the tail and head can be estimated from the early and late-time apparent resistivity. However, for a complex waveform, this method cannot be applied, and the solution we develop in this paper can only estimate the incomplete moments, which are more complex than for a step response.

The error in this estimation can be assessed by comparing the synthetic moment estimates with the analytic moment estimates provided by [30]. For this purpose, we present results for the AirTEM system [39] (shown in Figure 2), which is characterized by a triangular waveform similar to that of the AeroTEM system. This system employs a horizontal loop transmitter with a diameter of 4.25 m and a vertical magnetic coil receiver (dB/dt) located on the side of the transmitter loop. These are towed by a helicopter (Figure 2), and the EM system loop operates approximately 40 m above the ground. The transmitter waveform is a 1.85 ms bipolar triangular pulse with a frequency of 90 Hz. The time derivative of the magnetic field is sampled with 40 off-time channels between 0.02 and 3.3 ms and on-time differences covering the first half of each on-time pulse. The time domain synthetic half-space response is estimated using the program Airbeo [7] with Hankel filters [40] and frequency-to-time domain transformations [41]. The first two moments are estimated from the half-space response using the method proposed in this paper and are compared to the analytic solutions from [30], as seen in Figure 4.

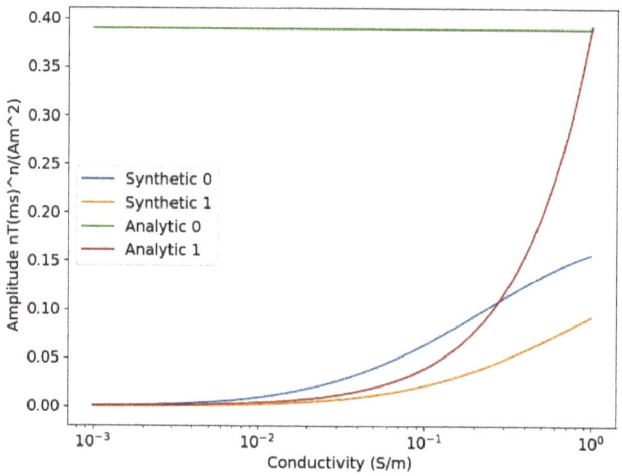

Figure 4. Comparison of synthetic and analytic half-space low-order moments.

2.5. Normalization

The moments of a wire loop were developed in [27], which illustrates the distribution of the moments for a simple electromagnetic model. For a wire loop with a time constant $\tau_w = L/R$, where L and R are the inductance and resistance, respectively, of the loop, the moments are

$$M^n = Bn!\tau_w^n \tag{28}$$

where B is the inductive limit and n is the moment order. This implies that time normalization will have an impact on the variation with the moment order. Accordingly, smaller time constants can be enhanced by scaling the moments. The moments of a scaled integral are developed as follows. If we define the scaled version of Y^n as Y_v^n, where

$$Y_v^n = \int_0^\infty y(t/v)t^n dt, \tag{29}$$

if $t/v = \tau$, this becomes

$$\begin{aligned} Y_v^n &= \int_0^\infty y(\tau)(v\tau)^n v d\tau \\ &= v^{n+1}\int_0^\infty y(\tau)\tau^n d\tau \\ &= v^{n+1}Y^n \end{aligned} \tag{30}$$

Scaling corresponds to multiplication of the moments by the power of a time factor. The impulse response moment from Equation (16) becomes

$$I_v^n = \frac{Y_v^{n+1} - \sum_{k=0}^{n-1}\begin{pmatrix} n+1 \\ k \end{pmatrix} X_v^{n+1-k} I_v^k}{\begin{pmatrix} n+1 \\ n \end{pmatrix} X_v^1} \tag{31}$$

Using a recurrent relation, it can be shown that

$$I_v^n = v^n I^n \tag{32}$$

This allows enhancing the moments of higher orders, if required.

2.6. Moment Estimation from Noisy Data

In practice, moments are estimated from noisy data, and an analysis of the proposed method requires an evaluation of the impact of noisy data on moment estimation. First, we must estimate the impacts of noisy data on the estimations of waveforms and output moments. These are estimated by numerical integration of the waveform and output signals, which can be written as

$$\begin{aligned} X^n &= \Sigma_0^l X(t_l) t_l^n \Delta t_l \\ Y^n &= \Sigma_0^m Y(t_m) t_m^n \Delta t_m \end{aligned} \tag{33}$$

where $X(t)$ and $Y(t)$ are the waveform and output signals, respectively, and t_l, t_l are time samples. Assuming independent measurements and neglecting correlations, the variances of these signals are estimated using the variance formula [42]:

$$s_f = \sqrt{\left(\frac{\partial f}{\partial x}\right)^2 s_x^2 + \left(\frac{\partial f}{\partial y}\right)^2 s_y^2 + \left(\frac{\partial f}{\partial z}\right)^2 s_z^2 + \ldots} \tag{34}$$

where s_f is the variance of the function $f = f(x, y, z, \ldots)$, and s_x, s_y, s_z are the variances of the parameters x, y, z. Developing the partial derivatives of Equation (33), we write

$$\begin{aligned} \frac{\partial X^n(t_l)}{\partial t_l} &= n t_l^{n-1} X(t_l) \Delta t_l \\ \frac{\partial Y^n(t_m)}{\partial t_m} &= n t_m^{n-1} Y(t_m) \Delta t_m \end{aligned} \tag{35}$$

Then, Equation (34) can be developed as

$$\begin{aligned} s_{X^n} &= \Sigma_0^l \left[n t_l^{n-1} X(t_l) \Delta t_l \right]^2 s_l^2 \\ s_{Y^n} &= \Sigma_0^m \left[n t_m^{n-1} Y(t_m) \Delta t_m \right]^2 s_m^2 \end{aligned} \tag{36}$$

where s_{X^n} and s_{Y^n} are the moment waveform and output variances respectively. Equation (36) shows that these moment variances increase with the moment order.

The impulse moment estimate variance can be developed in the following way. If we simplify Equation (19) for $n = 1$, we have

$$I^0 = \frac{Y^1}{X^1} \tag{37}$$

From Equation (34), we write

$$\begin{aligned} s_{I^0} &= \sqrt{\left(\frac{\partial I^0}{\partial Y^1}\right)^2 s_{Y^1}^2 + \left(\frac{\partial I^0}{\partial X^1}\right)^2 s_{X^1}^2} \\ s_{I^0} &= \sqrt{\frac{s_{Y1}^2}{(X^1)^2} + \frac{(Y^1)^2 s_{X_1}^2}{(X^1)^4}} \end{aligned} \tag{38}$$

In general, from Equation (19), the variance becomes

$$s_{I^n} = \sqrt{\left(\frac{\partial I^n}{\partial Y^{n+1}}\right)^2 s_{Y^n}^2 + \Sigma_{k=0}^{n-1} \left[\left(\frac{\partial I^n}{\partial I^k}\right)^2 s_{I^k}^2 \right] + \Sigma_{k=1}^{n} \left[\left(\frac{\partial I^n}{\partial X^k}\right)^2 s_{X^k}^2 \right]} \tag{39}$$

Without developing this expression in detail, it is easy to again conclude that the moment variances increase with the moment order.

3. Results

3.1. Synthetic Models

We apply this method on synthetic results from numerical models of a localized target in a conductive environment. We use the model of [9] to represent the sphere response to an AirTEM system in layered earth. The sphere has a radius of 50 m and a conductivity of 10 S/m, and its center is located 100 m below the surface of the ground, while the AirTEM system is at an altitude of 40 m. We start with a sphere in a half-space with 0.1 mS/m conductivity, for which the response is shown in Figure 5. The sphere has a distinctive moment response. The peak moments of a similar sphere with varying conductivity are displayed in Figure 6. This figure shows that the sphere's peak moments are negligible for conductivity below 1 S/m, even if the sphere has a meaningful off-time response, and the moment reaches a maximum value at about 8 S/m and decreases for conductivities above this value.

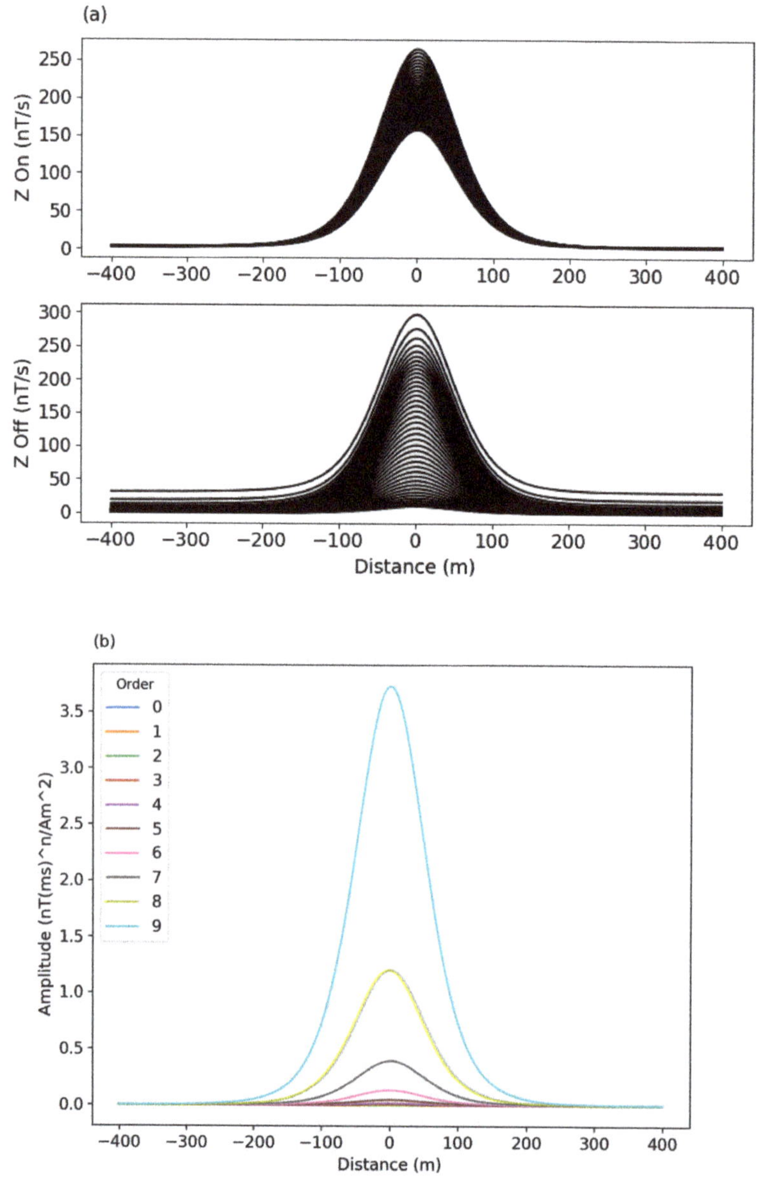

Figure 5. (**a**) On-time and off-time responses and (**b**) moments of a sphere with 10 S/m conductivity and a 50 m radius located in a conductive half-space of 0.1 mS/m 100 m below the surface.

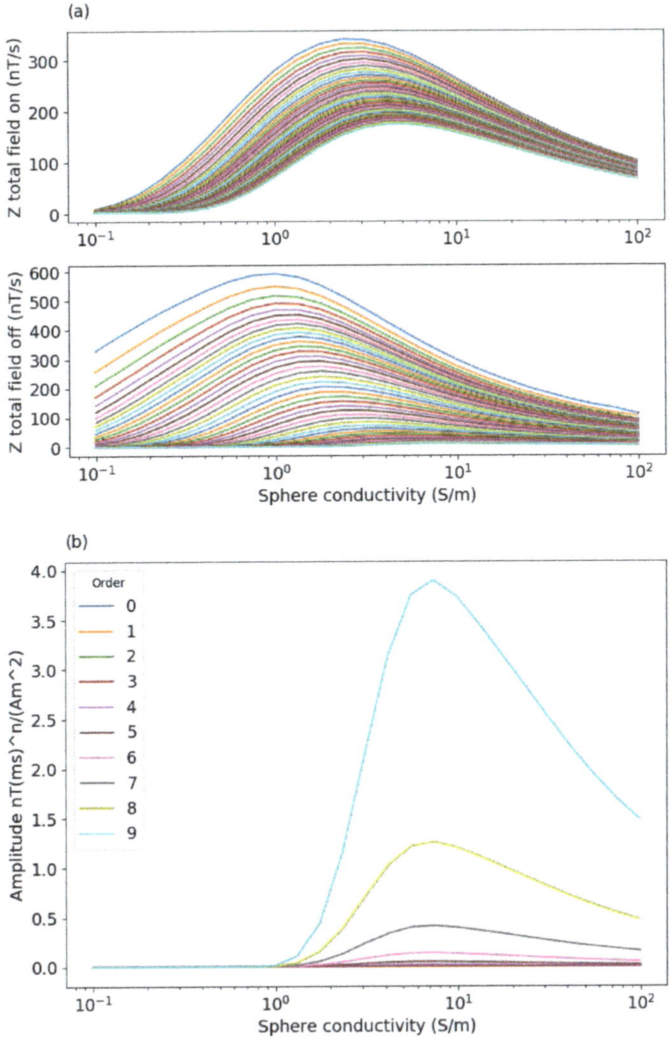

Figure 6. (**a**) On-time and off-time peak response nomogram and (**b**) associated moments for a sphere model with varying conductivity and 50 m radius located in a conductive half-space of 0.1 mS/m 100 m below the surface.

We now examine the sphere moments below a conductive overburden of 25 m thickness with varying conductivity. Figure 7 shows the peak nomogram for the overburden with varying conductivity ("anomalous" represents the sphere-only response). At low overburden conductivity, the response comes mainly from the sphere. As the overburden conductivity increases, the sphere peak becomes negative, while the overburden response increases. Figure 8 show the corresponding moments. At low overburden conductivity, the sphere moments dominate. However, above 0.1 S/m, the sphere moments decrease as the overburden moments increase strongly. Figure 8 suggests that we can still observe the sphere response if the overburden is 0.1 S/m (100 mS/m). This is confirmed by the subsequent model in Figure 9, wherein the top layer has a conductivity of 100 mS/m and

25 m thickness and is above a half-space with 0.1 mS/m conductivity. Even if the sphere's off-time responses are barely visible, the moment responses are clearly visible.

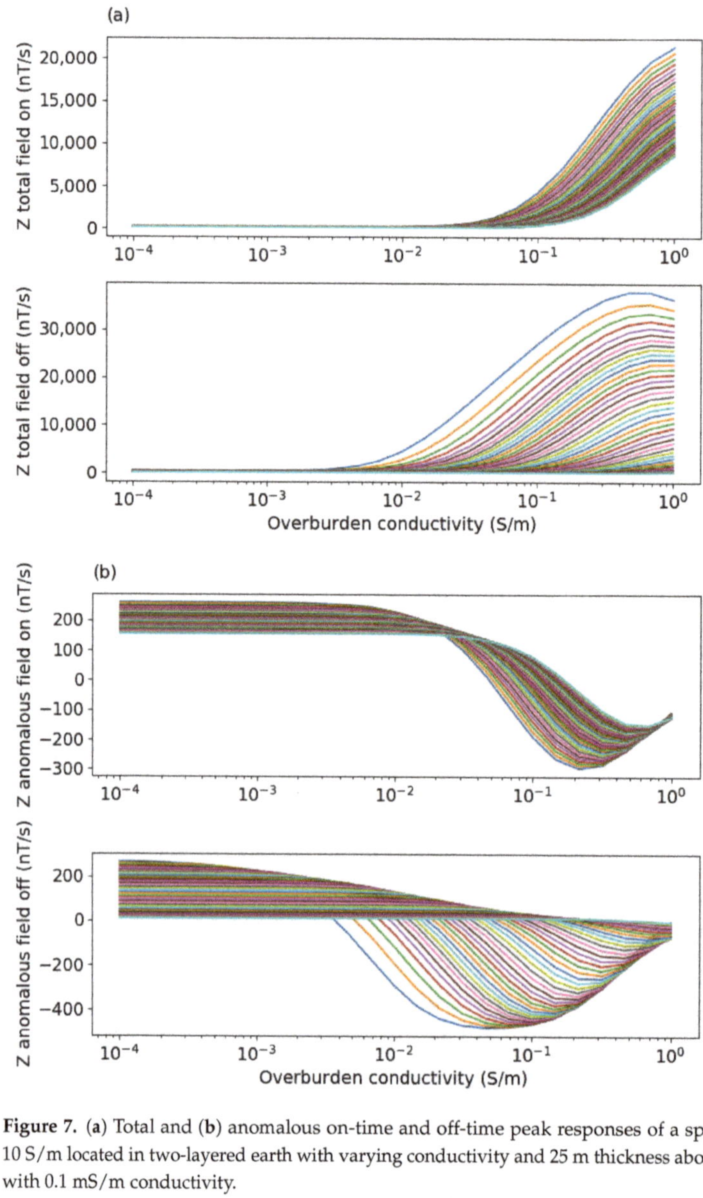

Figure 7. (**a**) Total and (**b**) anomalous on-time and off-time peak responses of a sphere model of 10 S/m located in two-layered earth with varying conductivity and 25 m thickness above a half-space with 0.1 mS/m conductivity.

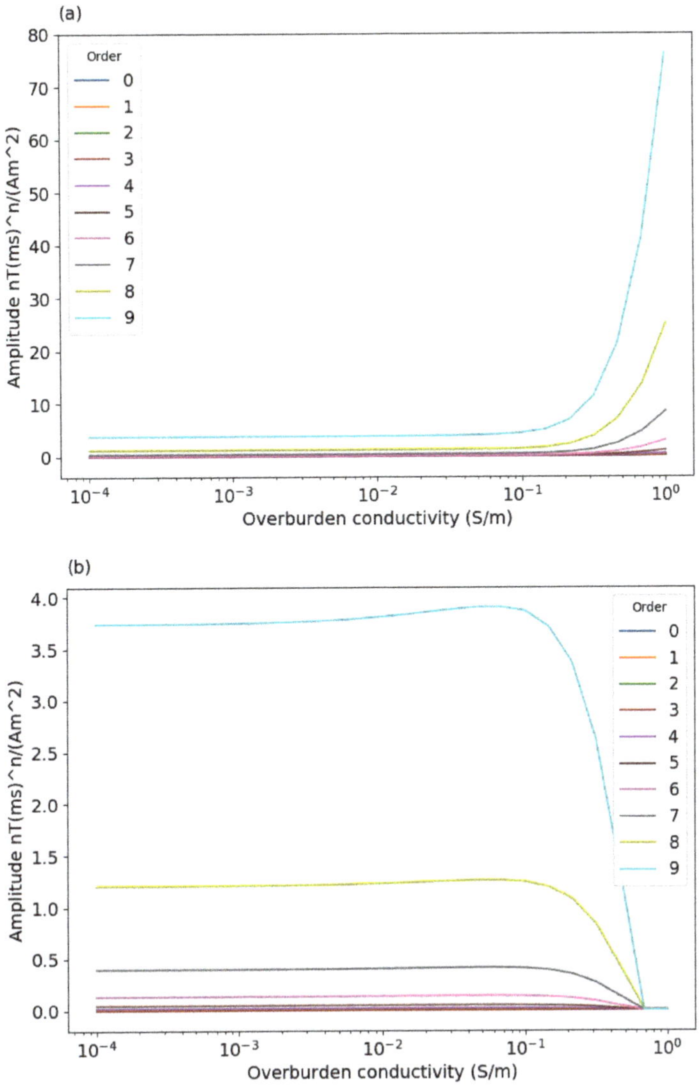

Figure 8. (**a**) Total and (**b**) anomalous peak response moments of a sphere model of 10 S/m located in two-layered earth with varying conductivity and 25 m thickness above a half-space with 0.1 mS/m conductivity.

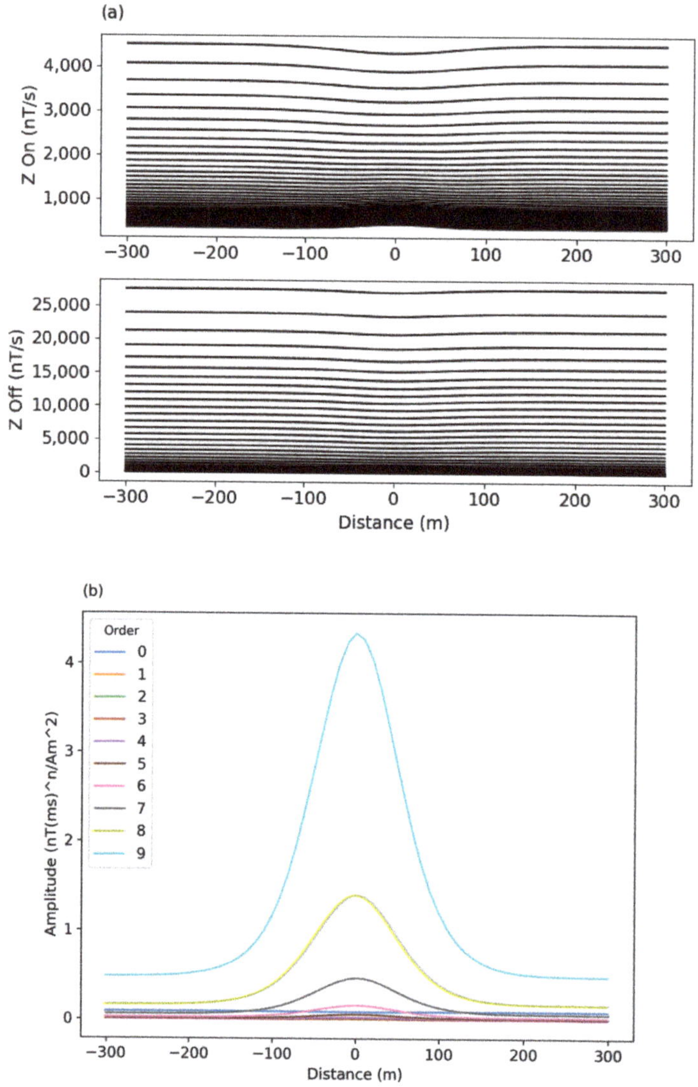

Figure 9. (**a**) On-time and off-time responses and (**b**) moments of a sphere model of 10 S/m, 50 m radius, and 100 m depth located in two-layered earth with 100 mS/m conductivity and 25 m thickness above a half-space with 0.1 mS/m conductivity.

3.2. Field Data

Moment estimation was tested on field data collected with the MEGATEM and the AirTEM systems over the Reid–Mahaffy test site in northern Ontario, Canada [43]. This site is characterized by the presence of several conductors located in a resistive basement below a variably conductive overburden.

3.2.1. MEGATEM

MEGATEM data collected over the Reid–Mahaffy test site in 2002 are available from [44]. In particular, the moments of Line 15, shown in Figure 10a, were analyzed by [27] based on B field estimation. For comparison, moments estimated using our proposed method are presented in Figure 10b. The various bedrock conductors can be clearly identified. Another profile of Line 3 in Figure 11 shows three distinct conductors whose moment signatures are remarkably significant.

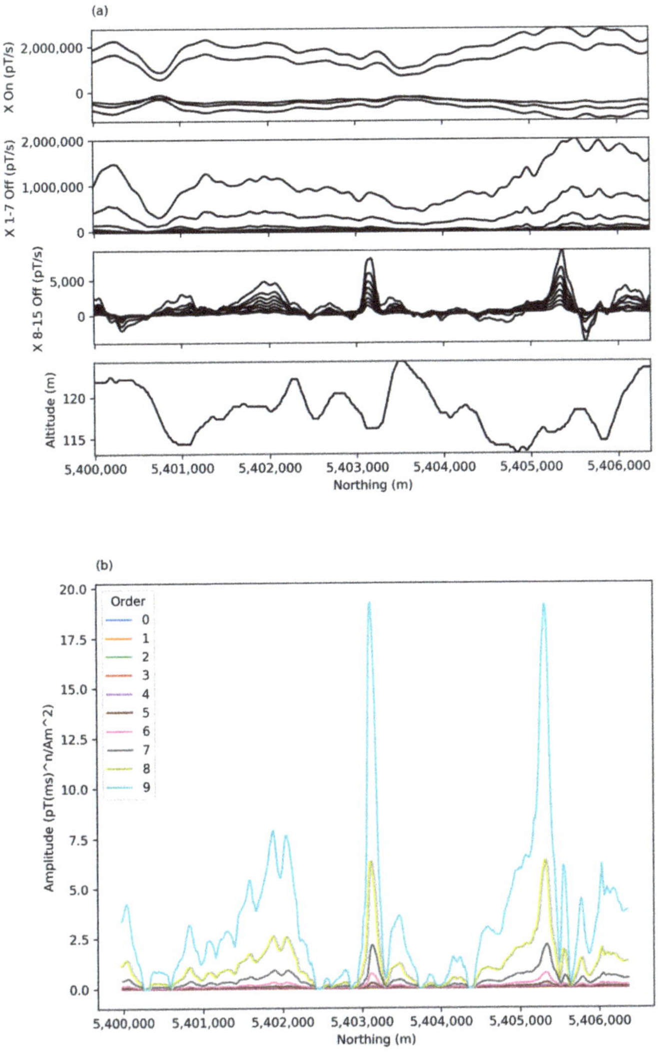

Figure 10. MEGATEM X component (**a**) on-time and off-time responses and (**b**) moments estimated from L15 of the Reid–Mahaffy test site.

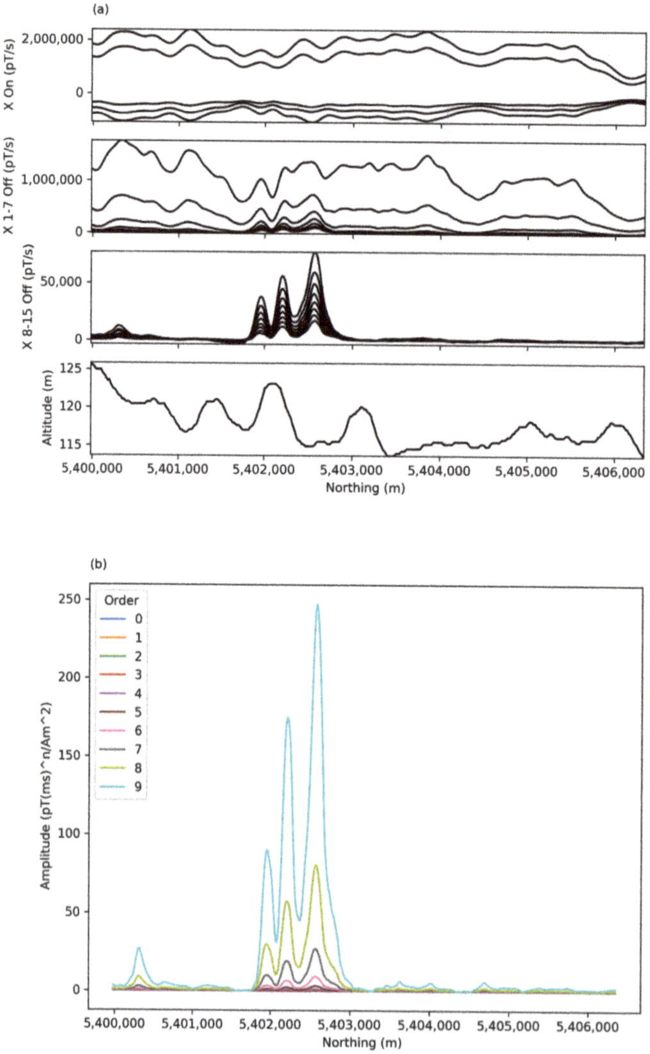

Figure 11. MEGATEM X component (**a**) on-time and off-time responses and (**b**) moments estimated from L3 of the Reid–Mahaffy test site.

3.2.2. AirTEM

AirTEM data were collected for testing purposes over the Reid–Mahaffy test site in 2019. The profiles for Line 15 are displayed in Figure 12a. AirTEM is a helicopter TDEM system with a much smaller loop flown closer to the ground. Consequently, the bedrock conductors are less apparent and are partly masked by the overburden response. Despite this fact, some deep conductors to the north of the profile are visible and can be observed in the moment response shown in Figure 12b. Line 3 conductor responses are shown in Figure 13. Clear moment responses are observed over these distinct conductors.

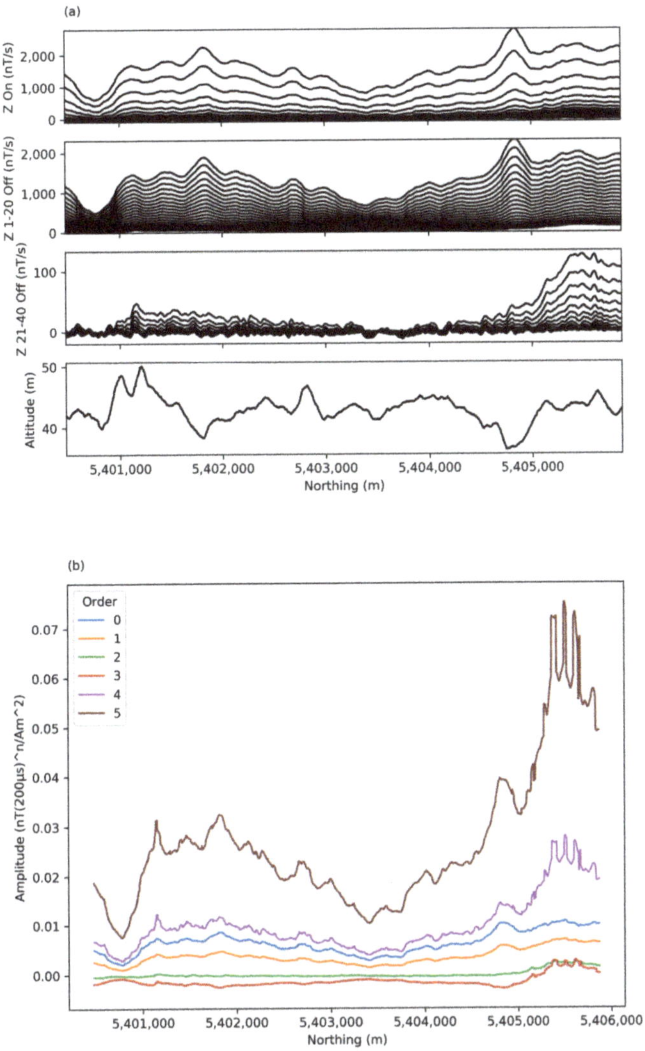

Figure 12. AirTEM (**a**) on-time and off-time responses and (**b**) estimated moments from L15 of the Reid–Mahaffy test site.

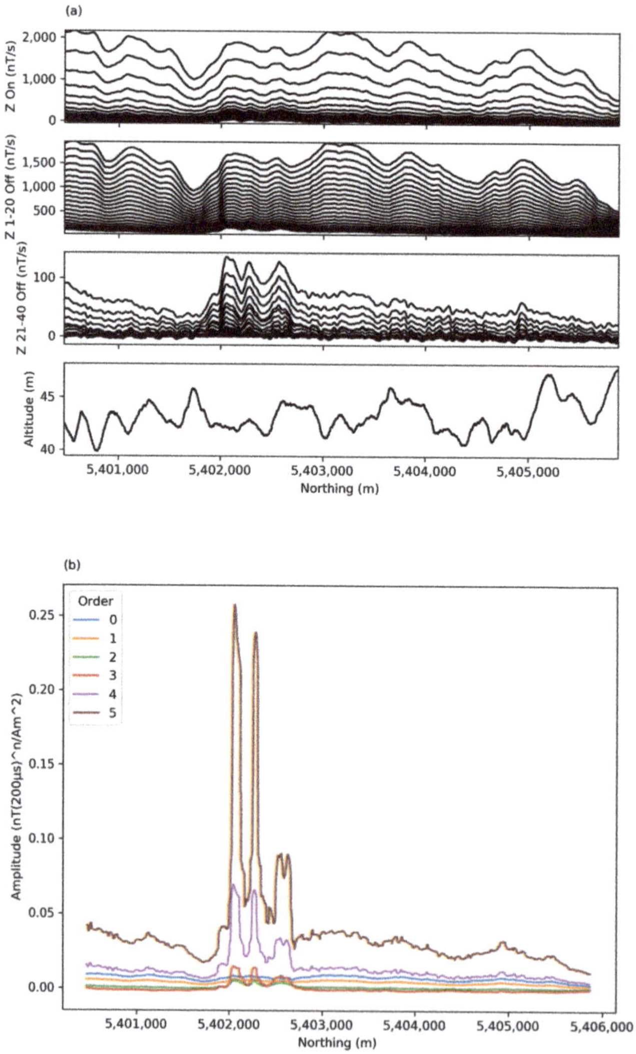

Figure 13. AirTEM (**a**) on-time and off-time responses and (**b**) estimated moments from L3 of the Reid–Mahaffy test site.

4. Discussion

Although not exhaustive, the results presented in this study allow a discussion of the strengths and limitations of the moment method. The methodology applied in this paper has been used successfully to interpret synthetic and field data. An important requirement of the method is the availability of on-time quadrature measurements. For the MEGATEM system, there is an established procedure for this purpose. For AeroTEM and AirTEM, some approximation is required. For other systems, this could be a major limitation to the implementation of this method.

The synthetic models show that for small background conductivity and a good conductor (in our models, more than a few S/m), the conductor moments are easily detected. However, as shown in Figure 6, a bad conductor (in our models, less than 1 S/m) will not be detected. Furthermore, when the conductivity of the background is significant, even a good conductor response is not detectable. Another important consideration when applying this technique is to consider the noise present in the data. The noise analysis shows that noise increases with the moment order, which may explain why high-order MEGATEM moments are less noisy than equivalent AirTEM moments, as the MEGATEM data, originating from a much more powerful system, are less noisy than the AirTEM data.

The application of this method on datasets from different systems over the same area allowed for a pertinent comparison of the performances of the systems in question. In the case presented, the Reid–Mahaffy area is covered with significant conductive overburden, and the AirTEM system, with smaller coincident loops, has some difficulty penetrating the overburden. Consequently, the discrete conductor moments are scattered. On the other hand, the MEGATEM system—operating at the same frequency but with much higher power, a larger transmitter loop, and greater transmitter–receiver separation—detects distinct conductors that are deeper below the overburden easily, comparatively speaking. These conductors are clearly defined by moments.

As mentioned in Section 2.4, the method presented in this paper provides an estimate of incomplete moments. This must be considered if results are compared with analytic models, and in that case [28], total moment relations must be applied. However, for numerical models wherein system time windows are included, observed moments can be directly related to model moments.

5. Conclusions

The method presented in this paper extends the application of moment estimation to arbitrary EM system waveforms and generalizes the use of moments; it can provide data quality assessments, conductor locations, and preliminary interpretations. A limitation of this approach is the availability of on-time measurements. For systems that can resolve this limitation, such as MEGATEM and AeroTEM (AirTEM), we show that this method successfully estimates moments and allows a comparison of system performance over a given area. We advocate that this method could be extended to other EM systems—either airborne, ground, or borehole—when on-time measurements are available. Our model also shows that for a conductor to be detected, the conductivity range must be appropriate: good conductors are easier to detect as long as there is contrast with the surrounding environment.

This algorithm can be easily implemented for data post-processing and can be used to glean information about the EM data quality quickly. As such, it is seen a supplement to simplified interpretation methods that only provide single parameters like decay time constants or apparent conductivities.

As this method relies on on-time and off-time measurements, improvements in data collection, like reducing noise and improving calibration methods, should improve the moment estimates. This is particularly true for on-time measurements: these are more difficult to collect, which explains why not all systems measure the full waveform.

Author Contributions: Conceptualization, M.A.V. and M.M.; methodology, M.A.V.; software, M.A.V.; validation, M.A.V. and M.M.; investigation, M.A.V.; resources, M.A.V.; writing—original draft preparation, M.A.V.; writing—review and editing, M.A.V. and M.M.; visualization, M.A.V.; supervision, M.M.; project administration, M.M. All authors have read and agreed to the published version of the manuscript.

Funding: This research received no external funding.

Data Availability Statement: MEGATEM data from Reid–Mahaffy from the given reference are available from the Ontario Geology website. AirTEM data are available on request from the first author.

Acknowledgments: The authors acknowledge Khorram Khan for revising the English.

Conflicts of Interest: The authors are both working for Geo Data Solutions GDS Inc., which commercializes the AirTEM system.

References

1. Grant, F.S.; West, G.F. *Interpretation Theory in Applied Geophysics*; McGraw-Hill Book Company: New York, NY, USA, 1965.
2. Ward, S.H.; Hohmann, G.W. Electromagnetic theory for geophysical applications. In *Electromagnetic Methods in Applied Geophysics: Volume I, Theory*; Nabighian, M.N., Ed.; Society of Exploration Geophysicists: Tulsa, OK, USA, 1987; pp. 131–311.
3. Kaufman, A.A.; Alekseev, D.; Oristaglio, M. *Principles of Electromagnetic Methods in Surface Geophysics*; Elsevier: Amsterdam, The Netherlands, 2014.
4. Palacky, G.J. Interpretation of INPUT AEM Measurements in Areas of Conductive Overburden. *Geophysics* **1975**, *40*, 490–502.
5. Keating, P.B.; Crossley, D.J. The inversion of time-domain airborne electromagnetic data using the plate model. *Geophysics* **1990**, *55*, 705–711.
6. Annan, A.P. The Equivalent Source Method for Electromagnetic Scattering Analysis and Its Geophysical Application. Ph.D. Thesis, Memorial University of Newfoundland, St. John's, NL, Canada, 1974.
7. Raiche, A.; Sugeng, F.; Wilson, G. Practical 3D AEM inversion—The P223F software suite. *ASEG Ext. Abstr.* **2007**, *1*, 1–5.
8. Vallée, M.A. New developments in AEM discrete conductor modelling and inversion. *Explor. Geophys.* **2015**, *46*, 97–111.
9. Vallée, M.A.; Moussaoui, M. Modelling the electromagnetic response of a sphere located in a layered earth. *Explor. Geophys.* **2023**, *54*, 362–375.
10. Hauser, J.; Gunning, J.; Annetts, D. Probabilistic inversion of airborne electromagnetic data for basement conductors. *Geophysics* **2016**, *81*, E389–E400.
11. Macnae, J.; King, A.; Stolz, N.; Osmakoff, A.; Blaha, A. Fast AEM data processing and inversion. *Explor. Geophys.* **1998**, *29*, 163–169.
12. Macnae, J. Automated airborne EM anomaly picking and 3D model fitting. *ASEG Ext. Abstr.* **2015**, *1*, 1–4.
13. Sattel, D.; Reid, J. Modelling of Airborne Em Anomalies with Magnetic and Electric Dipoles Buried in a Layered Earth. *Explor. Geophys.* **2006**, *37*, 254–260.
14. Kolaj, M.; Smith, R.S. Inductive electromagnetic data interpretation using a 3D distribution of 3D magnetic or electric dipoles. *Geophysics* **2017**, *82*, E187–E195.
15. Cox, L.H.; Wilson, G.A.; Zhdanov, M.S. 3D inversion of airborne electromagnetic data. *Geophysics* **2012**, *77*, WB59–WB69.
16. Cox, L.H.; Zhdanov, M.S.; Pitcher, D.H.; Niemi, J. Three-Dimensional Inversion of Induced Polarization Effects in Airborne Time Domain Electromagnetic Data Using the GEMTIP Model. *Minerals* **2023**, *13*, 779.
17. Cox, L.H.; Zhdanov, M.S.; Prikhodko, A. Inversion for 3D Conductivity and Chargeability Models Using EM Data Acquired by the New Airborne TargetEM System in Ontario, Canada. *Minerals* **2024**, *14*, 237.
18. Oldenburg, D.W.; Haber, E.; Shekhtman, R. Three dimensional inversion of multisource time domain electromagnetic data. *Geophysics* **2013**, *78*, E47–E57.
19. Ren, X.; Lai, M.; Wang, L.; Yin, C.; Liu, Y.; Su, Y.; Zhang, B.; Ben, F.; Huang, W. A fast 3-D inversion for airborne EM data using pre-conditioned stochastic gradient descent. *Geophys. J. Int.* **2023**, *234*, 737–754.
20. Fraser, D.C. Dighem resistivity techniques in airborne electro-magnetic mapping. In *Airborne Resistivity Mapping*; Palacky, G.J., Ed.; Geol. Surv. Canada Paper 86-22; Geological Survey of Canada: Ottawa, Canada, 1986; pp. 63–69.
21. Annan, A.P.; Smith, R.S.; Lemieux, J.; O'Connell, M.D.; Pedersen, R.N. Resistive-limit, time-domain AEM apparent conductivity. *Geophysics* **1996**, *61*, 93–99.
22. Nelson, P.H.; Morris, D.B. Theoretical response of a time-domain, airborne, electromagnetic system. *Geophysics* **1969**, *34*, 729–738.
23. Martinez, J.M.; Smith, R.; Diaz Vazquez, D. On the time decay constant of AEM systems: A semi-heuristic algorithm to validate calculations. *Explor. Geophys.* **2020**, *51*, 94–107.
24. Macnae, J.; King, A.; Stolz, N.; Klinkert, P. 3-D EM Inversion to the Limit. In *Three-Dimensional Electromagnetics*; Oristaglio, M., Spies, B., Eds.; Society of Exploration Geophysicists: Tulsa, OK, USA, 1999; pp. 489–501.
25. Schaa, R.; Fullagar, P.K. Rapid, approximate 3D inversion of transient electromagnetic (TEM) data. *SEG Tech. Program Expand. Abstr.* **2010**, 650–654.
26. Smith, R.S.; Lee, T.J. The impulse-response moments of a conductive sphere in a uniform field, a versatile and efficient electromagnetic model. *Explor. Geophys.* **2001**, *32*, 113–118.
27. Smith, R.S.; Lee, T.J. The moments of the impulse response: A new paradigm for the interpretation of transient electromagnetic data. *Geophysics* **2002**, *67*, 1095–1103.
28. Schaa, R.; Fullagar, P.K. Vertical and horizontal resistive limit formulas for a rectangular-loop source on a conductive half-space. *Geophysics* **2012**, *77*, E91–E99.
29. Fullagar, P.K.; Schaa, R. Analytic formulas for complete and incomplete first-order TEM moments below ground in a conductive half-space. *Geophysics* **2020**, *86*, E1–E11.
30. Smith, R.S.; Lee, T.J. Using the moments of a thick layer to map conductance and conductivity from airborne electromagnetic data. *J. Appl. Geoph.* **2002**, *49*, 173–183.
31. Lee, T.J.; Smith, R.S. Multiple-order moments of the transient electromagnetic response of a one-dimensional earth with finite conductance—Theory. *Explor. Geophys.* **2021**, *52*, 1–15.
32. Smith, R.; Hyde, C.; Lee, T.; Almond, R. Impulsive moments at work. *ASEG Ext. Abstr.* **2003**, *1*, 1–7.
33. Smith, R.S.; Fountain, D.; Allard, M. The MEGATEM fixed-wing transient EM system applied to mineral exploration: A discovery case history. *First Break* **2003**, *21*, 71–75.

34. Balch, S.J.; Boyko, W.P.; Paterson, N.R. The AeroTEM airborne electromagnetic system. *Lead. Edge* **2012**, *22*, 562–566.
35. Smith, R.S.; Lee, T.J. Multiple-order moments of the transient electromagnetic response of a one-dimensional earth with finite conductance—The Gaussian variation applied to a field example. *Explor. Geophys.* **2022**, *53*, 262–274.
36. Nash, J.E. Systematic Determination of Unit Hydrograph Parameters. *J. Geoph. Res.* **1959**, *64*, 111–115.
37. Koutsoyiannis, D.; Xanthopoulos, T. On the Parametric Approach to Unit Hydrograph Identification. *Water Resour. Manag.* **1989**, *3*, 107–128.
38. Smith, R.S. On removing the primary field from fixed-wing time-domain airborne electromagnetic data: Some consequences for quantitative modelling, estimating bird position and detecting perfect conductors. *Geophys. Prospect.* **2001**, *49*, 405–416.
39. Balch, S.J.; Samson, C.; Sanchez, J.L. Correcting for HTEM altitude variations by transmitter mutual inductance normalization. *SEG Tech. Program Expand. Abstr.* **2016**, 2139–2143.
40. Christensen, N.B. Optimized fast Hankel transform filters. *Geophys. Prospect.* **1990**, *38*, 545–568.
41. Raiche, A. Modelling the time-domain response of AEM systems. *Explor. Geophys.* **1998**, *29*, 103–106.
42. Ku, H.H. Notes on the use of propagation of error formulas. *J. Res. Natl. Bur. Stand.* **1966**, *70C*, 262.
43. Witherly, K.; Irvine, R.; Godbout, M. Reid Mahaffy Test Site, Ontario Canada: An example of benchmarking in airborne geophysics. *SEG Tech. Program Expand. Abstr.* **2004**, 1202–1204.
44. Ontario Geological Survey. *Ontario Airborne Geophysical Surveys, Magnetic and Electromagnetic Data, Timmins Area MEGATEM®*; Ontario Geological Survey, Geophysical Data Set 1041; Ontario Geological Survey: Sudbury, ON, Canada, 2002.

minerals

MDPI

Article

Location Prediction Study of Fluorite Ore in Shallow Cover Area: Evidence from Integrated Geophysical Surveys

Liu Cheng [1,2,3,*], Li Han [1,3], Yang Kai [1,3], Gao Yongbao [1,3], Tang Weidong [1,3] and Yao Chuan [1,3]

[1] Xi'an Center of Mineral Resources Survey, China Geological Survey, Xi'an 710100, China; dplihan@foxmail.com (L.H.); yangkaicgs@163.com (Y.K.); tangweidong23@sina.com (T.W.); yycc624@126.com (Y.C.)
[2] School of Geophysics and Information Technology, China University of Geosciences, Beijing 100083, China
[3] Technology Innovation Center for Gold Ore Exploration, China Geological Survey, Xi'an 710100, China
* Correspondence: liuchengxkd@email.cugb.edu.cn

Abstract: The Beishan region is a vital fluorite metallogenic belt in northwest China, characterized by favorable geological conditions for fluorite mineralization. However, being located in the Gobi Desert and affected by shallow cover layers, only a few outcrops can be observed on the surface. Therefore, comprehensive geophysical research is necessary to locate and predict regional metallogenic potential and the spatial distribution of veins beneath the cover. This study conducted a combination of ground magnetic method (GM), induced polarization (IP) surveys, portable gamma-ray (PGR), portable X-ray fluorescence analyses (PXRF), and audio Magnetotelluric (AMT) to conduct comprehensive exploration. The IP and GM effectively identified concealed ore-bearing space distributions and ground PGR- and PXRF-constrained mineralization anomalies, while AMT surveys constructed deep electrical structure models for ore deposits. This approach delineated concealed fluorite deposit locations as well as potential magmatic–hydrothermal migration pathways. Engineering verification confirmed the effectiveness of this method combination. This study established a comprehensive geological–geophysical positioning prediction technique that can serve as a reference for locating and predicting fluorite deposits in shallow-covered areas within the Gobi Desert.

Keywords: hidden fluorite ore; comprehensive geophysical method; Gobi Desert region; mineral exploration; Beishan metallogenic belt

1. Introduction

Currently, the majority of global mineral resources exploration efforts have shifted towards the deep continent and the covering layer. The traditional mineral exploration space is no longer sufficient to meet the increasing resource demands of humanity, prompting the need to broaden the search for new mineral spaces. As a result, mineral exploration has increasingly emphasized concealed exploration targets, with geophysics playing an increasingly vital role in the exploration process.

The China–Pacific region is a significant producer of fluorite, second only to Mexico in terms of reserves [1]. Currently, fluorite deposits in China are primarily exploited from surface and near-surface resources in the eastern region. However, as shallow fluorite resources are gradually depleted, future efforts are expected to focus on exploring fluorite deposits in the deeper and semi-arid to arid areas of the central and western parts of the country [2,3].

Drawing on the ample successful exploration experiences in covered terrains [4–7] and considering the current status of mineral resource development, it is evident that conducting mineral exploration and comprehensive prediction work in Cenozoic-covered areas such as Gobi Desert-covered areas is undeniably one of the best choices. Previous researchers have conducted theoretical studies and practical explorations on mineral deposits in covered terrains [8–12], primarily focusing on polymetallic deposits enriched with sulfides, where the mineralization location is intricately linked to the polarization parameters obtained from induced polarization methods. Additionally, based on density and magnetization anomalies, researchers have identified concealed rock body positions and forms using large-scale gravity and magnetic data to determine prospective exploration areas [13–15], showcasing a clear research direction. However, research on ore deposit prediction for concealed fluorite deposits still appears relatively weak. Given the insignificant differences in physical properties of fluorite and the limited scale of near-surface ore bodies typically exhibiting vein-like distributions, exploration techniques have predominantly focused on geochemical analysis and remote sensing interpretation [6,7,16,17]. Consequently, no notable geophysical exploration achievement currently targets concealed fluorite deposits. Deep-seated prediction of fluorite deposits is highly significant. Large to super-large fluorite deposits exhibit significant differences in deep-seated morphology compared to surface outcrops [18], suggesting substantial mineralization potential at depth. This underscores the necessity for penetrating geophysical methods to provide effective information regarding the occurrence of fluorite deposits beneath the Gobi Desert-covered area.

Due to the complexity of geological features, relying solely on individual geophysical, geochemical, or remote sensing anomalies makes it challenging to draw reliable conclusions in mineral resource prediction. In the case of epithermal vein-type fluorite deposits in the Gobi Desert shallow-covered areas, using integrated exploration methods primarily focusing on geophysical methods for ore body location prediction research is imperative. Moreover, characterizing the three-dimensional distribution of deep-seated fluorite deposits holds significant importance in the quest for concealed non-metallic sulfide mineral deposits.

2. Regional Geological and Geophysical Characteristics

2.1. Regional Geology

The research area is located in the eastern part of the Beishan metallogenic belt (refer to Figure 1). Positioned in the tectonic setting of the Tarim–Mongolia Plate, Siberian Plate, and Kazakhstan Plate intersection, the area lies within the middle section of the Tianshan–Xingmeng Orogenic Belt, a globally typical accretionary orogenic belt formed by subduction–collision processes from the early Neoproterozoic to Paleozoic eras [19–21]. The eastern segment of the Beishan region is characterized by a series of near-EW trending major faults, as well as NE trending faults [22]. Magmatic rocks of various ages generally exhibit northwest or near east–west-oriented band-like distributions consistent with regional tectonic lineaments. Fluorite deposits (points) in the region predominantly belong to the meso-low temperature magmatic–hydrothermal type, occurring in the area south of Huashitoushan–Gudongjing–Shahongshan. These deposits are mainly controlled by northeast, near north–south, and northwest-trending fault structures, displaying diverse host rocks, with the Yanshan period being predominant. Deposits such as the Jiaochigou fluorite deposit and the large tungsten–tin–rubidium fluorite deposit at Dongqiyi Mountain have been identified in the area, establishing it as a significant concentration zone for fluorite, tungsten, tin, rubidium, and beryllium mineralization within the Beishan metallogenic belt [23,24]. The study of the distribution patterns of fluorite deposits in the eastern Beishan area indicates a close association between the formation of fluorite deposits (points) and calcareous constructions, as well as intermediate-acidic volcanic or intrusive rocks. The northeast, near north–south, near east–west, and northwest-trending fault structures constitute the primary ore-hosting and ore-controlling structures for fluorite mineralization in the region, where the northeast and near north–south extensional structures are particularly favorable for mineralization. Moreover, the fluorite deposits in the area are predominantly hydrothermal infill-type deposits. Therefore, the detection of deep-seated fault structures and concealed rock bodies in the desert shallow-covered area is crucial for predicting the location and occurrence of structures that control hydrothermal solution movement within the region.

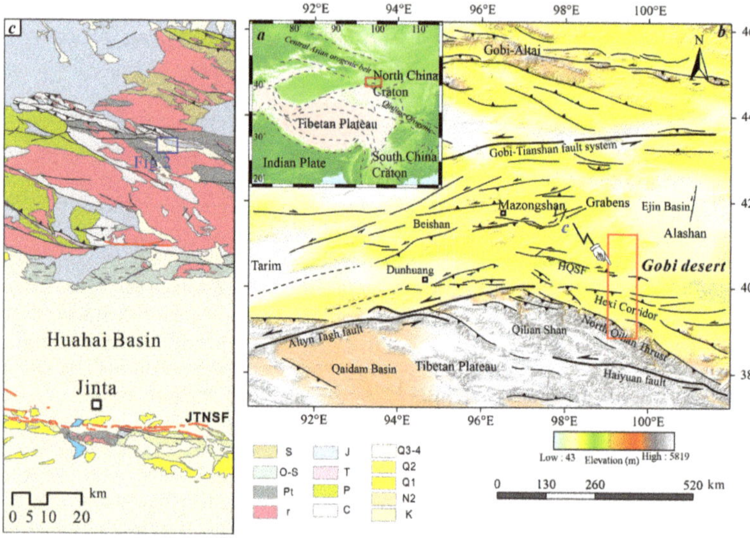

Figure 1. (**a**) Simplified tectonic subdivision map of the study area; the red box shows the study area. (**b**) Tectonic setting of eastern Tien Shan and Beishan Gobi Desert areas. (**c**) The geological map of the Beishan region, in which the blue box is the Gobi Desert shallow-covered area, is a hidden fluorite ore localization prediction technology test area, as shown in Figure 2.

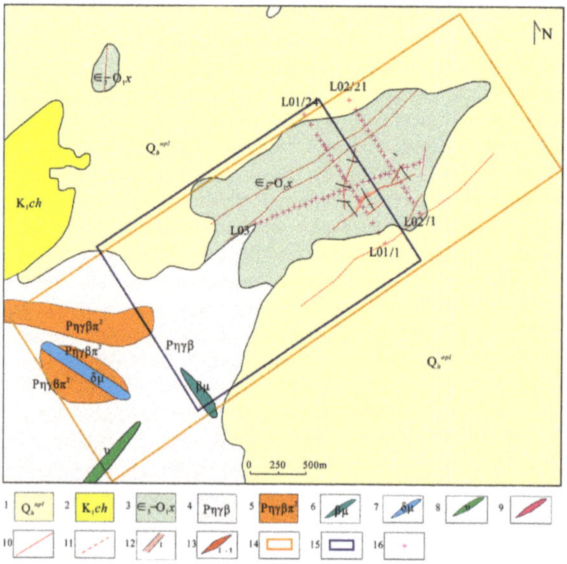

Figure 2. Mineral geological map of the study area and location of integrated geophysical exploration. 1. Quaternary gravel; 2. Lower Cretaceous Chijinbao Formation; 3. Upper Cambian–Lower Ordovician Xishuangyingshan Formation; 4. Late Permian biotite monzonitic granite; 5. Porphyritic biotite monzonitic granite; 6. Diabase dikes; 7. Diorite porphyrite veins; 8. Gabbro dikes; 9. Quartz veins; 10. Fault of unknown nature; 11. Inferred faults; 12. Location and number of mineralized alteration zones; 13. Location and number of ore bodies; 14. The working range of ground magnetic survey measurement is 15. The working range of the IP intermediate client method is 16. AMT sites.

2.2. Deposit Geology

The research area is situated on the northern side of the Xichangjing suture zone, at the southeastern margin of the early Paleozoic active continental margin of Gongpoquan–Dongqiyi Mountain, within the Gongpoquan–Dongqiyi Mountain metallogenic subzone. The fluorite deposits are found at the contact zone between rock bodies and strata, with the limited outcrop of strata in the area covered mainly by Quaternary deposits. In the northern part of the Suosuo Well rock body, the Cambrian–Ordovician West Shuangyingshan Formation strata are exposed, hosting ore structures that are lenses of sandstone interbedded with limestone within the West Shuangyingshan Formation. The area is characterized by a developed fault system, primarily trending northeast at around 55°, clearly visible on remote sensing images as a continuous negative topography. Additionally, a group of near-north–south-trending faults is also present, showing a lesser extent of extension [25]. The ore-controlling structures consist of two sets of northeast-trending faults forming a fault zone with strike-slip characteristics, with the near north–south-trending faults acting as secondary extensional faults sandwiched between the two northeast-trending faults. In the southern part, there are porphyritic biotite monzonitic granite rocks, with surrounding rocks showing alteration mainly characterized by hematization, limonitization, silicification, and gypsumization. The Suosuo Well rock body is exposed in the southern part, appearing as a rock mass outcrop, initially intrusive with fine-grained granite diorite in the late Silurian and later intruded by medium-grained porphyritic biotite monzonitic granite in the late Permian. The northeast-trending fault zone and the near evenly spaced north–south-trending fault structures between them collectively control the occurrence of structurally hydrothermal-type fluorite ore points at Huashitou Mountain (refer to Figure 1). The terrain in the area is relatively flat, with different degrees of surface covered by aeolian erosion material. Except for some exposed strata, the regional geological conditions are unclear, emphasizing the urgent need for penetrating methods to assess the mineralization potential and possible occurrence locations of fluorite deposits in the area.

2.3. Geophysical Characteristics of the Deposit

The predominant fluorite ore type in this area is the quartz–fluorite type, found in tectonic fault zones within rock bodies and formations without specific selectivity towards the surrounding rocks [26]. The surrounding rocks in the study area mainly consist of sandstone formations, with extensive Quaternary cover to the north and south. The extent of westward intrusion of rock bodies and the temporal–spatial relationships between the rock bodies and known mineral veins are not well defined. In several typical fluorite ore research areas, including this one, previous studies have indicated that fluorite ores exhibit minimal magnetism and typically manifest as negative magnetic anomalies in magnetic exploration results [27,28]. Quartz and ore-bearing quartz veins display weak magnetism, appearing as low magnetic anomalies (Table 1). If significant positive magnetic anomalies are generated by large-scale rock intrusions beneath the Quaternary cover in the study area, it could serve as compelling evidence for the source of fluorite mineralization and the mineralization mechanism. Statistical data on the physical properties of various fluorite ore regions indicate that, whether in magmatite, metamorphic, or sedimentary areas, the radioactive element content of fluorite is much lower than that of the host rock [29,30].

Table 1. The physical property parameters of the rocks and ores in the Huashitoushan working area.

Rock Types	Samples	Magnetic Susceptibility SI (10^{-6})	
		Range	Average
Altered cataclastic rock	10	1~5	3.34
Fluorite veins	15	0~5	1.06
Granite	10	36~179	81.3
Quartz sandstone	16	3~20	8.7

3. Integrated Geophysical Exploration and Processing

The majority of the study area is covered by varying degrees of aeolian erosion deposits, great surveys, and remote sensing technologies. Notably, the information provided by remote sensing techniques harbors numerous uncertainties, with their penetration capabilities in shallow-covered areas significantly inferior to geophysical exploration. Furthermore, the study area is situated in a desert environment characterized by a dry climate and high evaporation rates, leading to the formation of calcareous layers near the surface, posing challenges to mineral exploration primarily relying on geochemical methods. Although the fluorite ore exhibits limited differences in physical properties compared to the surrounding rocks [31,32], the geophysical targets that control hydrothermal solution movement are relatively well-defined based on their relationship with the deposit's location. Previous studies on existing fluorite deposits have indicated that fluorite mineralization shows no selectivity towards its surrounding rocks [7], predominantly occurring in several sets of parallel vein-like structures within breccias and shallow shear zones (faults) [26]. This provides explicit and detectable targets for deep-seated positioning predictions, making the corresponding geophysical detection targets more discernible. The desert mentioned above, with shallow cover, is an ideal area for experimenting with different geophysical exploration techniques.

Regarding mineral exploration, the most critical issue is not where minerals come from but rather the positions within the upper crust where minerals concentrate to form economic ore deposits [33,34]. The prediction and localization of fluorite deposits in shallow-covered desert areas based on comprehensive geophysical exploration mainly involves three levels of issues. Firstly, it is essential to determine potential occurrences of fluorite deposits within the region, explicitly identifying structures that control hydrothermal solution movement. Secondly, evaluating the mineralization potential of these structures is crucial. Lastly, locating and exploring the deep morphology and distribution of verified shallow fluorite veins to establish a reliable geological–geophysical model for guiding mineral development.

3.1. Ground Magnetic Method (GM)

As a rapid and cost-effective geophysical method, ground magnetic surveying can comprehensively reflect the magnetic anomalies of geological bodies to indicate the magnetic anomaly features of ore-controlling structures and fault zones, thereby revealing the presence and approximate distribution of structural mineralization alteration zones [35]. The magnetic properties of quartz and mineral veins are weak, resulting in low magnetic anomalies. This is especially evident in areas where host rocks comprise magmatic rocks, which display strong positive magnetic anomalies [28,36].

The magnetic survey data were obtained from multiple NW-oriented survey lines traversing the study area, with a spacing of 50 m between the magnetic survey lines. The spacing between magnetic survey points ranges from 5 to 20 m, densifying near known geological clues. Covering the entire study also expands the scope appropriately to the southwest-exposed biotite granite body to establish a sufficient background field. The magnetic survey employed the GSM-19T (GEM System, Canada) for magnetic measurements. Generally, the method of reduction to the pole is used to align the center of the magnetic anomaly to the top of the magnetic body. Reduction to the pole requires magnetic inclination, magnetic declination, and other geomagnetic parameters, which vary with latitude, longitude, and time.

The regional magnetic field distribution of the study area (Figure 3) is divided into four parts from west to east. Combined with the results of previous physical property parameter determinations (Table 1), the magnetism of the surrounding rocks, such as sandstone and quartz veins, is relatively weak, corresponding to low magnetic anomalies. Intrusive rocks exhibit distinct positive magnetic anomalies compared to fluorite and quartz. Although fluorite has the weakest magnetism due to its limited scale, it predominantly manifests as a composite response after being filled fault zone with hydrothermal fluids, resulting in a weak positive anomaly. The southwest side is characterized by a high-value positive anomaly area exceeding 100 nT. The high magnetic anomaly on the southwest side of the study area is inferred to be the manifestation of secondary intrusion of mafic rock veins such as diabase veins. Combining the magnetic parameter determinations of specimens, the more widely exposed biotite granite exhibits a weak positive anomaly. The eastern side features a relatively stable magnetic field and is also the primary region where hidden fluorite vein mineralization was discovered in this study. Considering surface outcrops, it is evident that the host rocks of fluorite ore are predominantly sandstone, which is weakly magnetic to non-magnetic. Fluorite ore is nearly non-magnetic, with a minimal difference in magnetism between the two. The magnetic anomalies of the fluorite and the quartz veins are weak in the positive anomaly transition zone.

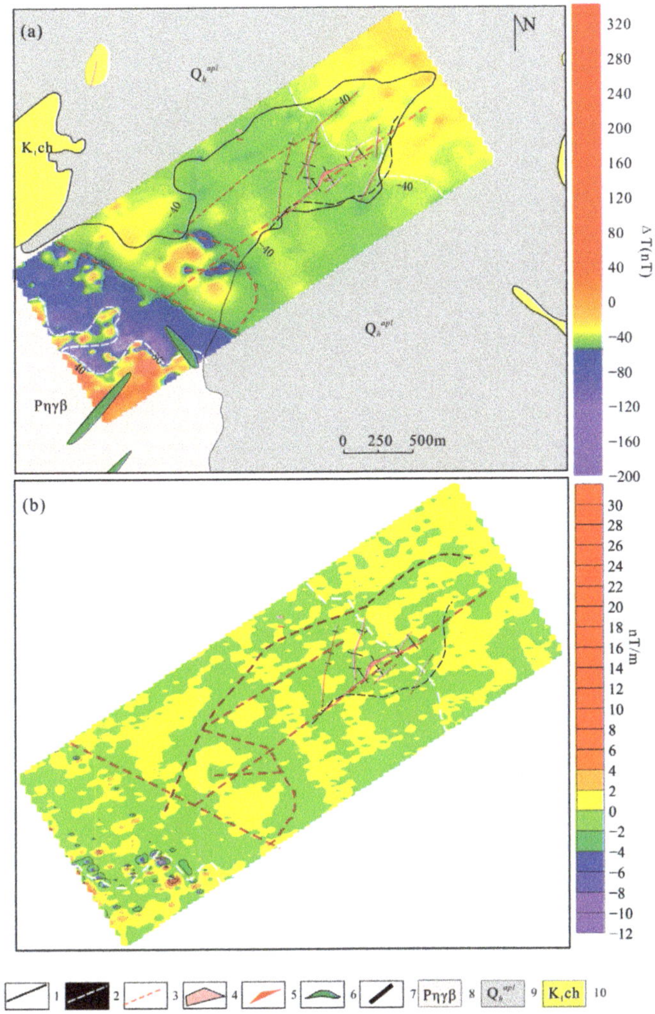

Figure 3. The map of reduction to the pole magnetic anomalies in the Huashitoushan, as well as the regional geological map (**a**) and contour of VFDMA (**b**). 1. Geological boundaries; 2. Prediction of geological boundaries; 3. Prediction of structural zones; 4. Mineralized alteration zone; 5. Fluorite ore body; 6. diabase veins; 7. Trenching position; 8. Late Permian biotite monzonitic granite; 9. Quaternary sand gravel; 10. Lower Cretaceous Chirinabe Formation.

The magnetic anomalies in the calm field and trench exploration indicate no traces of intrusive bodies on the south side of the critical working area of the Huashitoushan fluorite mine. Currently, the overall mining area appears to be situated within faults of sandstone formations. There is a noticeable difference in magnetic field values between the eastern side of the region and the verified mineralized area. Therefore, it is speculated that the location of the white line in the figure represents a geological interface. To the east of the white dashed line, hidden, weakly magnetic granitic rock masses may exist beneath the sandstone formations. The magnetic anomaly effectively indicates the extent of hidden fluorite mineralization within the area. In particular, the use of the vertical first derivative of the magnetic anomaly (VFDMA) modulus can better determine the horizontal projection position and distribution scale of anomalous bodies (Figure 3b). This method is particularly effective in identifying structures that control hydrothermal activity in areas with less pronounced magnetic contrasts.

3.2. Induced Polarization Method

The induced polarization method is primarily based on the low-resistivity anomalies of fluorite veins occurring within fault/fracture. Previous studies in fluorite mineral detection often utilized similar techniques, such as very low-frequency electromagnetic methods (VLF-EM) [37–41]. The shape and boundaries of various geological bodies can be ascertained by measuring their electrical resistivity. The resistivity of geological bodies is influenced by rocks' composition, structure, porosity, and water content. In contrast to intact dense rocks, fault zones usually have lower resistivity because of higher water content from groundwater evaporation or substantial mineral/sulfide enrichments. Hence, the induced polarization method can effectively identify weak areas such as faults and other structural weaknesses (Table 2).

Table 2. The electrical parameters of the rocks and ores in the Huashitoushan working area.

Rock Types	Samples	ρ (Ω·m)			η (%)		
		Minimum	Maximum	Average	Minimum	Maximum	Average
Quartz sandstone	5	4722	9398	7037	0.895	2.276	1.74
Quartz vein	5	1978	10,712	5598	0.306	1.426	0.77
Tectonic breccia	5	5403.3	5403.3	5403.3	1.02	2.82	1.82
Altered cataclastic rock	5	6822.0	16,830.5	11,223.3	0.75	1.70	1.12
quaternary	20	3	52	18.72			

The induced polarization survey coincided with GM surveys with a spacing of 20 m. The resistivity anomalies in the study area vividly delineate the resistivity structure. In Figure 4, significant low-resistivity anomalies are observed along the survey lines on both the north and south sides. Combined with the trench exploration revealing the absence of salt–alkali weathering shell shielding, the anomalies comprehensively reflect the Quaternary sand–gravel cover layer. The red dashed line on the southern side marks a distinctive resistivity gradient zone, representing not only the rapid thinning of desert cover but also corresponding to a hidden structural zone, verified by the trench exploration as the main fluorite ore-forming fault within the area. Within the central part of the profile, a series of low-resistivity anomalies oriented in a northeast direction appear in the sandstone formation. A wide low-resistivity belt extending eastward on the northern side, nearly a hundred meters wide and still open-ended, likely reflecting earlier geological work identifying the main fault. However, the induced polarization work indicates the main fault direction lies to the south, with the identified fault here being a secondary parallel fault. Apart from the primary NE-oriented fault, the presence of an NNE-oriented hidden secondary fault inferred from low-resistivity anomalies was also confirmed by trench exploration. Given the relatively flat terrain of the desert region, the resistivity parameters obtained from the induced polarization method are minimally affected by topography, providing a genuine reflection of underground resistivity structure features at specific depths. The method proves to be sensitive in detecting hidden faults exhibiting low-resistivity anomalies.

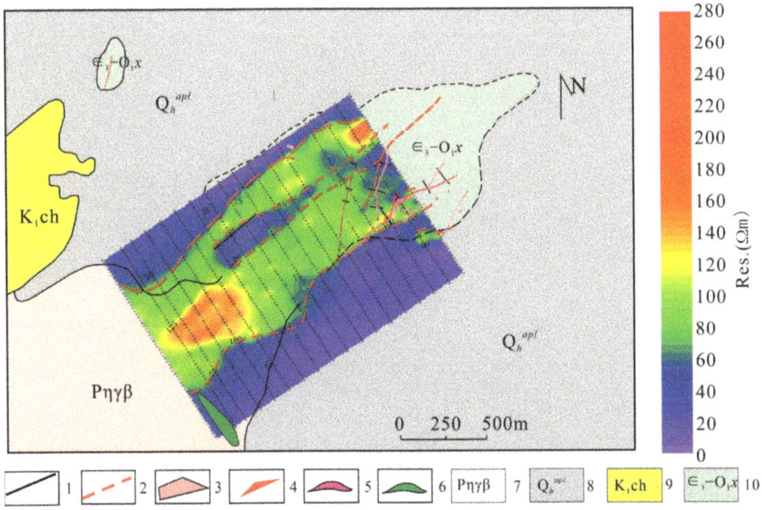

Figure 4. Induced polarization (IP) resistivity anomaly diagram of the detection area in the Huashitoushan and the regional geological map. 1. Geological boundaries; 2. Prediction of structural zones; 3. Mineralized alteration zone; 4. Fluorite ore body; 5. Quartz veins; 6. Diabase dikes; 7. Late Permian biotite monzonitic granite; 8. Quaternary sand gravel; 9. Lower Cretaceous Chijinbao Formation; 10. Upper Cambian–Lower Ordovician West Shuangyingshan Formation.

3.3. Ore-Bearing Potential Detection Technology of Fluorite Ore in Shallow Covering Area

Portable gamma-ray (PGR) measurements infer the position of concealed ore bodies by measuring the overall variation trend in radioactive elements within different geological formations. In evaluating the mineral potential of fluorite deposits, ore-bearing structural zones hinder the upward migration of radioactive elements due to vein fillings, resulting in differences in radioactivity between these locations and faults with open channels. Additionally, the radioactivity of vein-like fluorite minerals is generally lower than that of the surrounding rock properties, whether the surrounding rock is sedimentary strata or rock mass (Table 3). Based on other technical means, such as detecting hidden faults in shallow-covered areas of the Gobi Desert, rapid judgment of the metallogenic potential of favorable locations can be made after initial favorable positions for ore formation exploration [29,30].

Portable X-ray fluorescence analyses (PXRF) determine element types and their respective concentrations based on characteristic X-rays emitted from samples; however, this technique is currently limited to detecting Ca-related elements closely associated with fluorite and cannot analyze F-element content directly [16,43]. Ground gamma-ray spectrometry measurements and portable X-ray fluorescence analysis complement existing ground high-precision magnetic surveys and induced polarization (IP) surveys. Soil samples are collected from beneath a 10 cm layer of surface erosion material to avoid interference from Quaternary gravel cover layers. For exposed bedrock or thin covering layers, measurement points are adjusted to ensure uniformity in measurement media.

Table 3. Statistics of lithologic radioactivity content in some fluorite deposits (modified from [30,42]).

Lithology	Tc/10^{-6}		Deposit Location
	Variation Interval	Mean Value	
granite	35.4~99.3	65.9	Southern Songxian County, Henan Province, China
Rhyolite porphyry	25.6~70.3	50.1	
Altered granite	40.6~95.1	62.3	
Altered rhyolite	21.0~72.3	48.9	
Massive fluorite ore	11.2~40.0	23.0	
Cemented fluorite ore	13.0~48.3	25.8	
Banded fluorite ore	14.2~45.5	27.3	
Quartz–fluorite ore	15.8~50.4	26.5	
Fine vein fluorite ore	14.1~73.4	37.6	
sand slate	9.4~25.0	15.9	Linxi County, Inner Mongolia, China
fluorite ore	4.5~8.9	6.3	
limestone	6.7~20.0	11.03	Pengshui County, Chongqing, China
fluorite ore	0.9~7.4	3.86	

3.4. Audio Magnetotelluric Sounding

The methods mentioned above are primarily focused on predicting the planar distribution of concealed fluorite deposits. Due to the close relationship between the spatial occurrence of hydrothermal solution movement and fault structures, it is challenging to assess their occurrences and variations in deep-seated areas using conventional surface methods. Furthermore, within expanded spaces of fractured zones at depth, there is a tendency for significant enlargement of high-grade fluorite ore bodies [6]. Electromagnetic exploration methods play an irreplaceable role in mineral exploration by utilizing high-frequency electromagnetic signals to probe underground electrical structures, providing abundant electrical information for exploring deep-seated concealed mineral deposits and complex structural regions [44–46]. In the Gobi Desert region, where artificial electromagnetic interference is minimal, obtaining good data can be easily achieved by effectively reducing ground resistance. The resistivity distribution obtained through inversion can be translated into lithological units related to rock resistivity.

3.4.1. Data Acquisition and Processing

The audio Magnetotelluric method is an effective method for studying underground resistivity structures. In 2022–2023, the Institute of Xi'an Center of Mineral Resources Survey conducted audio Magnetotelluric surveys at 73 points in the Huashitoushan of the Beishan metallogenic belt, with a distance of approximately 20–40 m. The Crystal Global Aether instrument was used. At each measuring point, two horizontal electric field components (Ex and Ey) and two magnetic field components (Hx, Hy) were recorded for more than 0.5 h with a frequency band range of 10k Hz~5Hz, which corresponds roughly to the depth range of 0~2 km in a homogeneous half-space with a resistivity of 100 Ω·m. The study area exhibits favorable terrain conditions and low levels of interference. However, certain regions are affected by surface aridity, with some even being covered by desert. Consequently, the grounding conditions during data collection were suboptimal. To address this issue, deep excavation followed by clay filling and watering, deep burial, and moisture retention through plastic film coverage can effectively reduce ground resistance and enhance data reliability. We use the prMT software (v1.0.3.5) to process the collected audio Magnetotelluric data. The collected Magnetotelluric data were converted from the time domain to the frequency domain through a standard robust algorithm [47]. The impedance tensor was obtained, and the power spectrum was selected interactively for better results. Thanks to the relatively low electromagnetic interference in the Gobi Desert shallow-covered area, most of the data quality at most stations is excellent (Figure S1), with only a few points (L03P18) showing higher-than-normal resistivity values in the high-frequency range due to arid surface conditions.

3.4.2. Two-Dimensional Inversion and Results

This paper adopts the two-dimensional NLCG algorithm [48] for data inversion. Theoretically, the joint inversion mode of TE+TM can yield optimal results; however, the strict requirement of two-dimensionality for subsurface media limits the applicability of TE mode [49,50], thus making TM mode data the primary basis for inversion. Phase data are subject to minimal distortion, with an assumed error of 20% in apparent resistivity and 10% in phase for TM mode, while for TE mode, errors are set at 80% in apparent resistivity and 60% in phase. The frequency range during the inversion process spans from 5 to 10k Hz, with a regularization factor τ set at 3. The final fitting error after inversion is determined as 0.89.

Based on the completed vital profile of the two-dimensional electrical structure, it is evident that the central part of the profile exhibits a series of alternating high and low resistivity anomalies. Combining these results with induced polarization (IP) surveys, multiple significant faults within the sandstone formations, labeled F1–F9 from south to north, have been identified. Among these faults, F1 and F9 represent the sandstone formation's southern and northern boundaries, respectively. The low-resistivity anomalies within 100 m near-surface depth at both ends in the north–south direction are attributed to Quaternary cover layers. This confirms that large-scale low-resistivity anomalies observed during IP surveys are not caused by shallow saline–alkali layers or massive, intrusive bodies beneath desert cover layers in the south. It further negates the possibility of extensive rock intrusion below desert cover layers in the south. Magmatic–hydrothermal activity related to regional mineralization is likely sourced from deep-seated F2 faults or high-resistivity anomalies on this section's northern side.

Audio Magnetotelluric (AMT) results indicate numerous low-resistivity anomalies associated with fault zones provide ample space for fluorite mineralization. Faults such as F2, F3, F4, and F6 have all been exposed through trench exploration activities, confirming their existence as concealed fault zones. Additionally, known faults such as F7 and F8 established during previous geological work align with electric resistivity models obtained through two-dimensional inversion techniques. The known vein locations extend to considerable depths underground, reaching approximately 400 m below ground level; meanwhile, two high-resistivity anomalies appearing in both central and northern parts below a depth of 400 m are speculated to reflect intermediate-acidic rock formations within this region, which may serve as sources for material and energy contributing to regional mineralization.

3.4.3. Three-Dimensional Inversion and Results

With the widespread adoption of the ModEM program [51,52], there has been a significant reduction in memory requirements and an improvement in inversion speed. As a result, three-dimensional inversion techniques for Magnetotelluric (MT) data have gradually gained wide application in scientific research. In recent years, advancements in algorithm technology and computational capabilities have led to three-dimensional inversion replacing two-dimensional inversion as the mainstream MT inversion technique.

The inversion grid design is as follows: the unit grid widths in the x and y directions are 25 m, with nine grids extended in the x and y directions and an extension step size of 1.5 times. The total number of inversion grids is 61 (x direction) $\times$ 67 (y direction). The vertical depth of the first layer is 10 m, with a layer thickness increment factor of 1.1. There are five expanded grids, an external expansion factor of 1.5, with 40 grids total, and the bottom grid node is located at 12 km z-direction). The final number of AMT measurement points used in the inversion calculation is 73, with 22 frequencies utilized in the inversion process ranging from 5 to 10k Hz. The data selected for inversion consists of the off-diagonal apparent resistivity components (ρxy and ρyx) and phase (θxy and θyx). In the inversion, we set a lower error limit of 5% for the Zxy and Zyx impedance components. In terms of inversion parameter settings, the regularization factor is adaptively reduced in the ModEM program. Therefore, during the inversion process, we set its initial value to

1000, and when the change in fitting difference is less than 2×10^{-3}, the value is updated by dividing by 10. When the regularization factor is less than 10^{-8}, or the fitting difference is less than 1.05, the inversion is stopped. After 84 iterations, nRMS reached 1.86 (The sensitivity test process and corresponding results are shown in Supplementary Documents and Figures S3 and S4).

The horizontal slices of the 3D resistivity model (Figure 5) reveal a general low-resistivity characteristic beneath the shallow cover layers in the Gobi Desert, with deep electrical structures primarily composed of high-resistivity sandstone formations and associated fault structures. Cross-sections at the location of each AMT line of 3D inversion (Figure S2) are similar to 2D inversion results (Figure 6); both exhibit electrical, structural features of relatively high-resistivity sandstone horizons divided by several sets of low-resistivity anomaly faults. The high-resistivity anomalies R1 and R2 represent the sandstone formations beneath the shallow cover layers, while the intervening low-resistivity anomalies correspond to the fault system. The low-resistivity anomalies C2 and those on the southern side of the study area collectively form a main structural trend-oriented NEE (northeast–east), whereas C1 and C3 low-resistivity anomalies constitute a set of nearly NE-trending faults. Given that there are no magnetic anomalies, as previously mentioned, it is inferred that continuous high-resistivity anomalies R1 and R2 represent intact sandstone formations within this region. Additionally, in the eastern parts of the study area, high-resistivity anomaly R3 exhibits weak positive magnetic anomaly characteristics likely related to intrusive rocks.

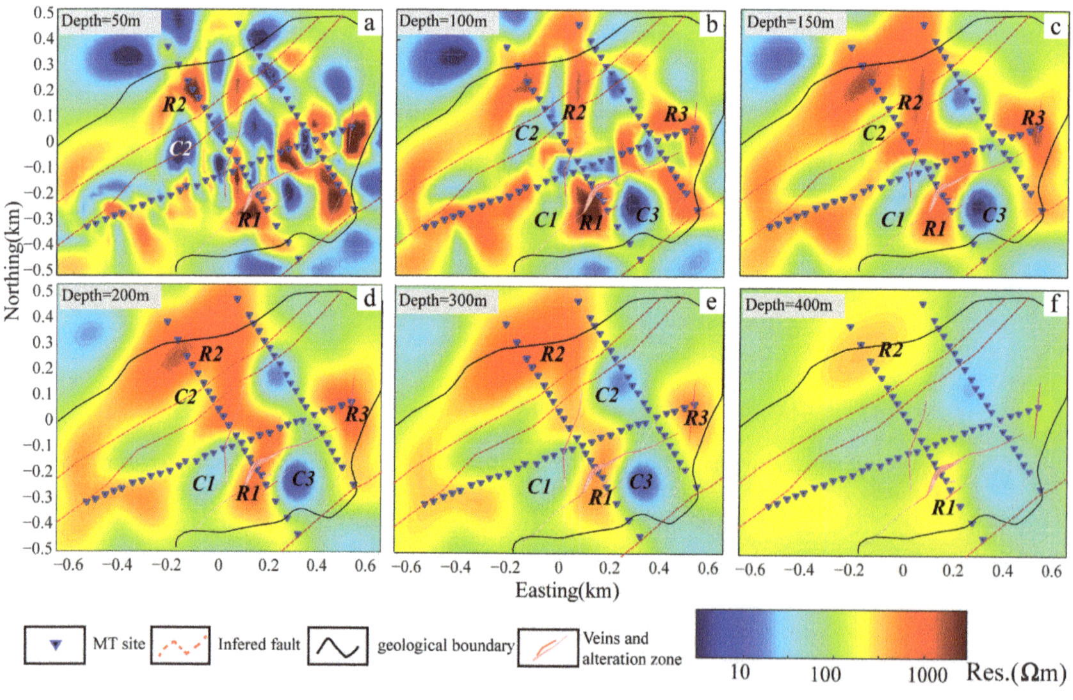

Figure 5. Horizontal slices of the 3D inversion model at depths of 50, 100, 150, 200, 300, and 400 m (**a**–**f**), respectively. The superposed geological features are the same as in Figure 2. Warm and cold colors indicate high and low resistivity, respectively.

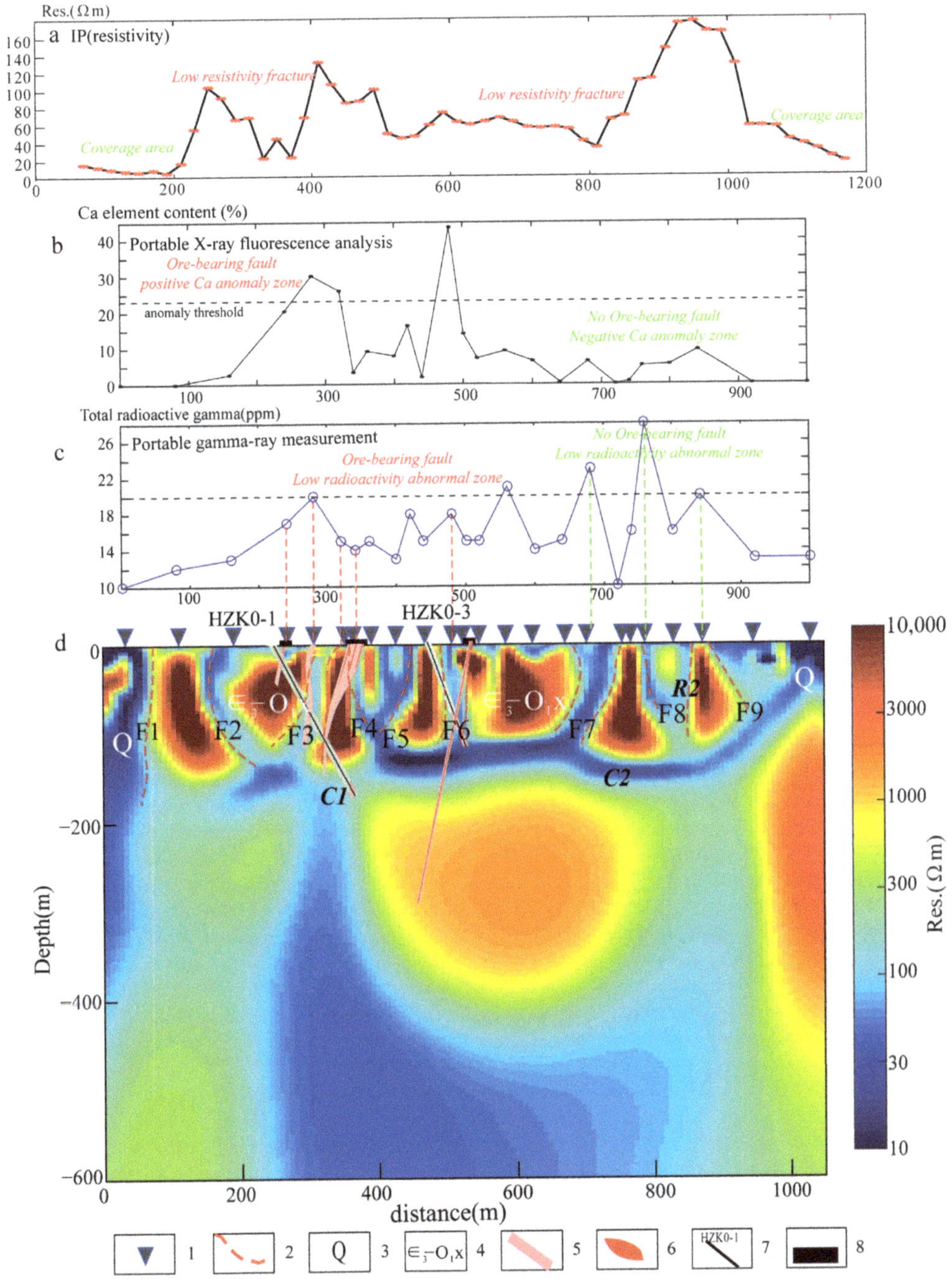

Figure 6. Line 1 integrated geophysical profile ((**a**) IP resistivity; (**b**) PXRF; (**c**) PGR; (**d**) AMT 2d inversion).

4. Discussion

4.1. Constraints of Integrated Geophysical Exploration Results on Hidden Fluorite Deposits

Combining the more sensitive AMT two-dimensional inversion resistivity model for shallow details [53] with the previously mentioned fluorite ore potential detection technology, the central part of the profile displays a series of alternating high and low resistivity anomalies. In conjunction with induced polarization (IP) survey results, a series of significant faults, labeled F1–F9, are observed from south to north within the sandstone formations. F1 and F9 represent the southern and northern boundaries of the sandstone formations, with low resistivity anomalies within 100 m of the surface at the south and north ends, indicating Quaternary cover layers. This confirms that the large-scale low resistivity anomalies observed during the IP surveys are not due to shallow saline–alkali layers, further negating the possibility of large-scale intrusive rock formations beneath the desert cover layer to the south. Magmatic–hydrothermal activity related to regional mineralization is likely to originate from the deep parts of the F2 fault or the high resistivity anomalies to the north of the profile. The AMT results indicate that the multiple low resistivity anomalies represent fault zones with abundant fluorite mineralization potential. The presence of concealed fault zones has been verified by the exposure of the F2, F3, F4, and F6 faults through trenching. F7 and F8 correspond to known faults delineated in previous geological work, consistent with the resistivity model obtained from the two-dimensional inversion. It is currently known that mineral veins extend to depth, with mineralization occurring at depths of approximately 200 m underground. Two high-resistivity anomalies in the central and northern parts of the study area below a depth of 200 m are speculated to reflect intermediate-acidic rock formations within the region, providing material and energy sources for regional metallogenic.

The metallogenic potential of the faults mentioned above varies, and the zonation of fluorite ore bodies is determined by the process of ore-bearing hydrothermal fluids rising along specific negative pressure structures and filling fractured spaces. The magmatic activity provides energy and material sources, while secondary faults and dense fissures within the surrounding rock provide space for the aggregation of ore-forming hydrothermal fluids and material exchange. Geophysical surveys utilizing TDIP and GM methods can effectively locate favorable mineralized positions. However, uniformly distributed mineralization is extremely rare, and the selective distribution of ore bodies still requires further constraint through the combination of PGR and PXRF. PGR can rapidly obtain near-surface radioactivity. The gamma radiation in the soil originates from the decay of K, U, and Th elements in deep geological bodies [54], with the fundamental rule being that volcanic rocks exhibit higher radioactivity than sedimentary and metamorphic rocks [29]. The mineralization process of fluorite is related to volcanic rocks' energy and material sources. As fluorite ore and its associated quartz veins are present within structural faults, this hinders the upward migration of deep-seated radioactivity. Consequently, high-value anomalies correspond to polymetallic alteration in the surrounding rock, while low-value anomalies correspond to quartz veins and fluorite ore bodies.

The results from profile line 0 indicate that the total radioactivity levels at the measurement points near the F3, F4, and F6 faults associated with mineralization are all below the lower limit of 20 ppm, while the non-mineralized fault zones F7, F8, and F9 exhibit high anomalies. However, the F1 and F2 faults where the cover is thicker also show low total radioactivity levels, making it challenging to make judgments using a single method. It is imperative to integrate geological conditions with other methods to ensure the accuracy of interpretation. Although portable X-ray fluorescence analysis cannot directly detect the content of fluorine (F) elements, it can directly determine the calcium (Ca) elements closely related to fluorite minerals on-site, avoiding the lengthy chemical analysis cycle and providing constraint conditions rapidly. The data show that the Ca element content at measurement points near F3, F4, and F6 is all above 20%, while the Ca element content at measurement points near fault zones known not to have fluorite mineralization is below 10%. Therefore, high Ca element values can effectively indicate the possibility of fluorite mineralization.

Based on the deployment of drilling verification in conjunction with integrated geophysical exploration for positioning and prediction, the preliminary control of the ore body ranges from 100 to 657 m in length, with a thickness of 0.71 to 1.74 m and an inclination depth down to 47.88 m, exhibiting CaF_2 grades ranging from 15.64% to 60.64%. A comparison between surface and deep borehole mineral characteristics reveals a trend in varying thickness, gradual enrichment, and increasing richness within the fluorite deposits at depth, consistent with the well-matched spatial distribution revealed by AMT resistivity inversion models (Figure 7). Currently, verified mineral deposit locations almost correspond to the boundaries of high resistivity anomalies, displaying a solid spatial correlation with intact sandstone formations. Conversely, no more significant fluorite deposits have been found within large-scale low-resistivity anomalies. The application of 3D inversion provides a basis for the different occurrences of mineral deposits in various regions of the ore body. Preliminary estimates indicate approximately 109,800 tons of fluorite mineral resources within depths shallower than 50 m, demonstrating significant resource potential. The discovery of fluorite deposits at Huashitou Mountain—particularly the constraints imposed by comprehensive geophysical exploration results on deep-seated mineralization space and its associated state—holds valuable implications for prospecting similar genetically related fluorite deposits in shallow-covered areas such as the Gobi Desert.

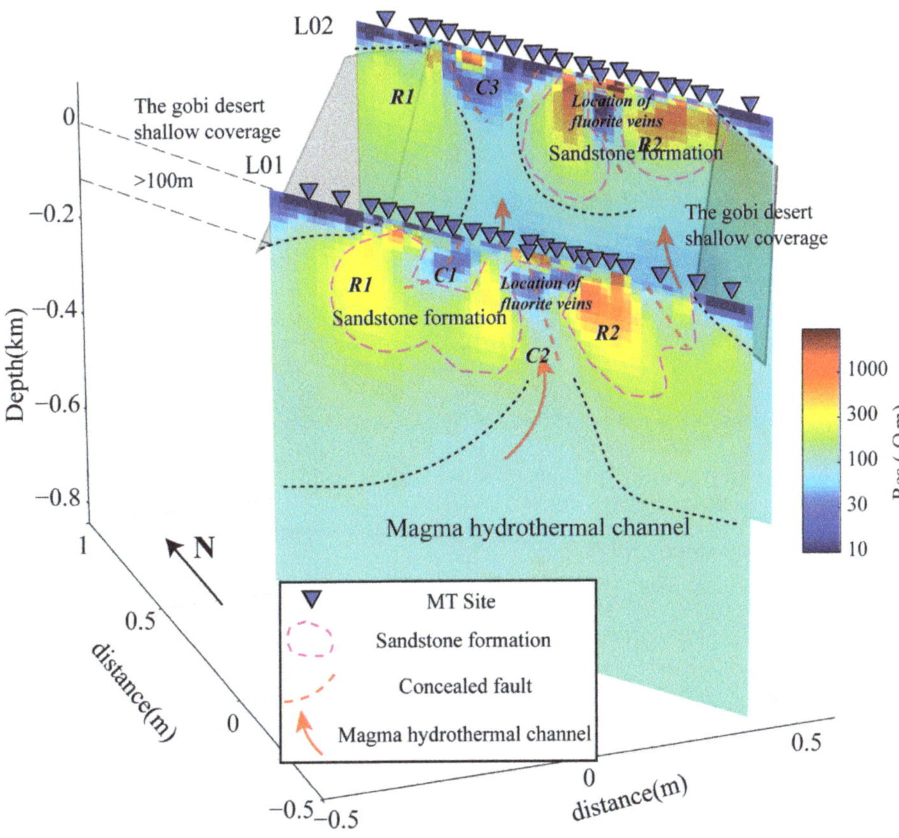

Figure 7. Comparison diagram of the AMT comprehensive profile.

4.2. Method System of Shallow Coverage Area Positioning Prediction Technology

Various geophysical exploration methods possess distinct characteristics, application limitations, and corresponding conditions. When conducting work in the shallow-covered areas of the Gobi Desert, the shielding effect of near-surface cover layers impacts all technical methods to a certain extent. Therefore, it is essential to explore the optimal combination of techniques for exploring concealed fluorite deposits and locating their positions primarily controlled by faults.

This study employed GM surveys and induced polarization (IP) to detect structures that control hydrothermal solution movement. Previous research also incorporated high-resolution multispectral remote sensing, or VLF. While satellite remote sensing can rapidly identify large-scale fault locations and dimensions within a region—serving as an indispensable tool for preliminary positioning predictions—it is crucial to rely on penetrating geophysical methods for detecting faults that may host fluorite deposits due to favorable passage conditions in the Beishan metallogenic belt. The effectiveness of GM surveys exhibits selectivity for surrounding rocks hosting fluorite deposits; it performs better when magmatic rocks are present but shows limited differentiation in sedimentary rock environments. The completed IP survey effectively reflects potential fault structures hosting fluorite deposits, compared to VLF commonly used in previous studies, which exhibit weaker signal strength and less prominent anomalies than the IP method [55,56]. Although the low association between fluorite deposits and sulfides results in insignificant polarization anomalies, IP remains an effective prospecting method, especially in topographically stable regions such as the Gobi Desert. Overall, considering these factors will aid in determining suitable combinations of geophysical techniques for exploring concealed mineral resources under similar geological settings.

Detecting the mineral potential of fluorite deposits is a challenging research task. Based on the physical properties discussed earlier, fluorite deposits are relatively small in scale and exhibit limited variations in physical properties. The utilization of PGR and PXRF can enhance prediction precision. Both methods are influenced by the thickness of the surface cover, as evidenced by the stable low anomalies of both the ground gamma-ray total counts and the Ca element content in the heavily covered regions on the north and south sides of the study area, as shown in the figure above. PGR requires comprehensive assessment with the inferred fault positions from other methods. The factors causing anomalies are complex, but they hold unique indicative significance for the occurrence of hydrothermal solution movement. PXRF can effectively provide evidence of the Ca content, providing direct evidence for the mineral potential in concealed faults. However, it is worth noting that external alluvial sediments or drift loads in high-calcium background areas necessitate comprehensive analysis combined with geological surveys and other technologies.

Although a set of convenient and effective geophysical methods for the shallow-covered desert areas has been summarized through comprehensive geophysical surveys, the inherent ambiguity in the inversion and interpretation of individual methods often leads to conflicting results in integrated interpretations. This issue is particularly pronounced when dealing with non-metallic sulfide deposits such as fluorite, where there is a weak correlation between the occurrence location and mineralization potential. By utilizing a joint inversion and interpretation approach integrating seismic velocity structure with magnetic, electrical, and radiometric data, an organically unified geological–geophysical model with mutually constrained physical parameters, including magnetism, resistivity, and radioactivity, can more effectively predict the location of concealed fluorite deposits.

4.3. Formation Mechanism of Mineral Deposits

By combining the GM, TDIP, PGR, and AMT, we studied the geophysical–geologic structure of the study area from shallow to deep layers to identify potential ore targets in the Huashitoushan area. The lithology in the drilling confirms the results. Further analysis of the deposit's formation mechanism will support additional work. In particular, the AMT three-dimensional inversion resistivity model provides a geophysical basis for imaging the deep formation of the mineralization space of the deposit in the near-surface small-scale sporadic distribution and the deep morphology of the deep metallogenic material transportation channels within the deposit (Figure 8).

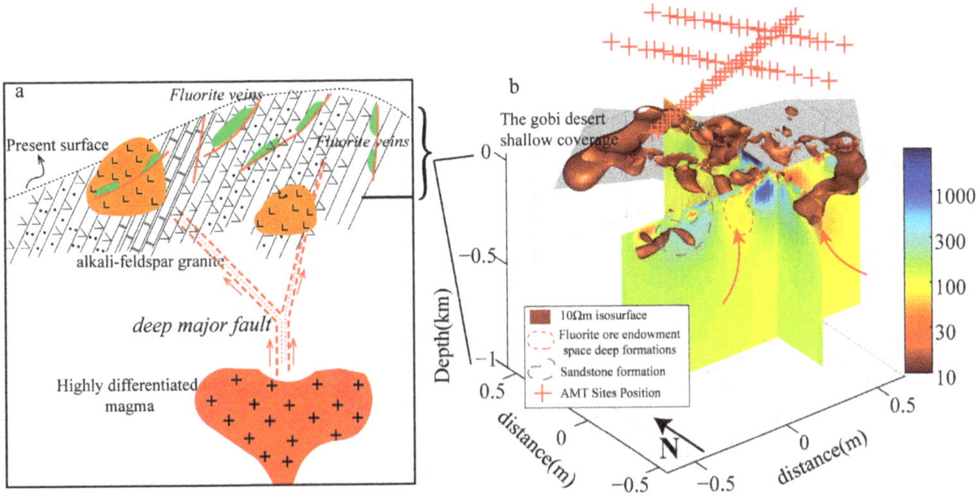

Figure 8. An integrated sketch diagram illustrating possible formation mechanisms of fluorite deposit in the Beishan metallogenic belt (**a**) and a perspective display of the 3D resistivity model and schematic geological interpretation (**b**).

Fluorite is primarily composed of F and Ca, exhibiting characteristics of in situ mineralization. Based on this, it is believed that the fluorite deposit at Huashitou Mountain may have originated mainly from medium-grained porphyritic biotite monzogranites for F and sandstone formations for Ca. In the eastern part of Beishan, a certain depth of crustal material melting exists to form an acidic magma chamber. During differentiation and evolution, fluoride-rich magmatic–hydrothermal fluids intrude along deep-seated faults and encounter relatively "cooler" atmospheric precipitation within the identified fault system through comprehensive geophysical detection. In shallow, near-surface areas, fluoride-rich fluids undergo water–rock reactions with surrounding rocks, continuously extracting Ca elements from the surrounding rock mass to enrich precipitates, forming early-stage fluorite deposits when temperature decreases further while pH increases [31,57]. The mineral fluid accumulates in favorable spaces (fault zones, cavities), forming fluorite deposits. The distribution of fluorite deposits within the mining area strictly follows control by faults and silicified fault zones, displaying characteristics typical of fill-type deposits. Fluorite deposition occurs under conditions where magma has fully differentiated and calcium and volatile components have been sufficiently enriched; these conditions lead to mineral precipitation within sandstone formations along northeast-trending or secondary north–south-trending fault structures.

5. Conclusions

Ground magnetic method, IP intermediate gradient scanning, audio Magnetotelluric, portable gamma-ray, and portable X-ray fluorescence analyses were carried out successively in the Huashitoushan fluorite deposit and Beishan metallogenic belt, and the different working methods attained several achievements.

1. Summarized the phased fluorite deposit positioning and prediction technology process in the shallow-covered areas of the Gobi Desert. Initially, GM and IP surveys were utilized to identify potential concealed ore-bearing fault structures, representing

possible structures that control hydrothermal solution movement. Subsequently, PGR and PXRF were employed to constrain anomalies associated with potential mineralization within spatial occurrences of fluorite deposits. Finally, combining deep-seated information provided by electromagnetic detection with drilling and trenching verification led to discovering multiple concealed fluorite deposits, providing valuable references for positioning and predicting fluorite deposits in shallow-covered areas.

2. The resistivity model constrained by the audio Magnetotelluric method, especially 3D inversion, was used to constrain the deep subsurface structures of known mineralized faults. The fluorite ore bodies of Huashitoushan are mainly located at the junction regions of high-resistivity sandstone formation and relatively high-conductivity fault/fissure alteration zones.
3. Unlike the traditional geophysical detection of metal–sulfide deposits in shallow coverage areas, the physical properties of the ore body and the surrounding rocks are significantly different, and the prediction of fluorite ore positioning needs to be constrained by the combination of two phases of spatial detection of endowment and metallogenetic potential, as well as by a combination of various technological methods.

Supplementary Materials: The following supporting information can be downloaded at: https://www.mdpi.com/article/10.3390/min14080838/s1, Figure S1: Observed apparent resistivity-phase curves for four representative AMT stations (L01P08, L01P19, L02P04 and L03P18); Figure S2. cross-sections at the location of each AMT line of 3D inversion; Figure S3. The sensitivity test of the resistivity structure corresponds to the example shown along the L01 and L02 profiles in the 3D resistivity model. The resistivity of the fixed block in the initial model was replaced at 0–200 m; Figure S4. shows the response curve after the forward simulation of the sensitivity test compared with the originally proposed curve.

Author Contributions: Conceptualization, L.C.; methodology, Y.K. and L.H.; software, L.C. and L.H.; validation, L.H., T.W. and Y.C.; investigation, L.C., Y.K. and T.W.; resources, G.Y. and T.W.; data curation, L.C.; writing—original draft preparation, L.C.; writing—review and editing, G.Y.; visualization, Y.K. and L.H.; supervision, G.Y.; project administration, L.C. and T.W.; funding acquisition, L.C. All authors have read and agreed to the published version of the manuscript.

Funding: This research was funded by the Natural Science Basic Research Program of Shaanxi (grant No. 2023-JC-QN-0365) and the China Geological Survey Project (grant Nos. DD20243342, DD20243333, DD20211552, and DD20240019).

Data Availability Statement: The model and data files of 3D inversion were compressed and uploaded as Supplementary Files.

Acknowledgments: We thank Gary Egbert and his group members for sharing their 3D inversion code. We thank Alan Jones and Gary NcNeice for using their tensor decomposition codes. We thank Dong Hao for using their EM3dvp code.

Conflicts of Interest: The authors declare no conflicts of interest.

References

1. Summaries, M.C. *Mineral Commodity Summaries;* US Geological Survey: Reston, VA, USA, 2021; p. 200.
2. Camprubí, A.; González-Partida, E.; Richard, A.; Boiron, M.-C.; González-Ruiz, L.E.; Aguilar-Ramírez, C.F.; Fuentes-Guzmán, E.; González-Ruiz, D.; Legouix, C. MVT-like fluorite deposits and oligocene magmatic-hydrothermal fluorite–Be–U–Mo–P–V overprints in northern Coahuila, Mexico. *Minerals* **2019**, *9*, 58. [CrossRef]
3. Castorina, F.; Masi, U.; Gorello, I. Rare earth element and Sr-Nd isotopic evidence for the origin of fluorite from the Silius vein deposit (southeastern Sardinia, Italy). *J. Geochem. Explor.* **2020**, *215*, 106535. [CrossRef]
4. Hosseini-Dinani, H.; Aftabi, A. Vertical lithogeochemical halos and zoning vectors at Goushfil Zn–Pb deposit, Irankuh district, southwestern Isfahan, Iran: Implications for concealed ore exploration and genetic models. *Ore Geol. Rev.* **2016**, *72*, 1004–1021. [CrossRef]
5. Shah, A.K.; Bedrosian, P.A.; Anderson, E.D.; Kelley, K.D.; Lang, J. Integrated geophysical imaging of a concealed mineral deposit: A case study of the world-class Pebble porphyry deposit in southwestern Alaska. *Geophysics* **2013**, *78*, B317–B328. [CrossRef]

6. Tang, L.; Zhang, S.; Wang, L.; Pei, Q.; Fang, Y.; Cao, H.; Zou, H.; Yin, S. Exploration of concealed fluorite deposit in shallow overburden areas: A case study in Elimutai, Inner Mongolia, China. *Earth Sci. Front.* **2021**, *28*, 208.
7. Zou, H.; Pei, Q.; Li, X.; Zhang, S.; Ware, B.; Zhang, Q.; Fang, Y.; Yu, H. Application of field-portable geophysical and geochemical methods for tracing the Mesozoic-Cenozoic vein-type fluorite deposits in shallow overburden areas: A case from the Wuliji'Oboo deposit, Inner Mongolia, NE China. *Ore Geol. Rev.* **2022**, *142*, 104685. [CrossRef]
8. Schamper, C.; Rejiba, F.; Guérin, R. 1D single-site and laterally constrained inversion of multifrequency and multicomponent ground-based electromagnetic induction data—Application to the investigation of a near-surface clayey overburden. *Geophysics* **2012**, *77*, WB19–WB35. [CrossRef]
9. Meng, Y.; Zhang, R.; Liu, R.; Lin, T.; Wang, W. Application of Multiple Geophysical Methods to Prospect Concealed Ores Beneath Quaternary Cover: A Case Study from a Copper-Polymetallic Deposit. In *Technology and Application of Environmental and Engineering Geophysics, Proceedings of the 7th International Conference on Environmental and Engineering Geophysics, ICEEG, Beijing, China, 26–29 June 2016*; Di, Q., Xue, G., Xia, J., Eds.; Springer: Singapore, 2017; pp. 243–249.
10. Xue, G.; Zhang, L.; Hou, D.; Liu, H.; Ding, Y.; Wang, C.; Luo, X. Integrated geological and geophysical investigations for the discovery of deeply buried gold–polymetallic deposits in China. *Geol. J.* **2020**, *55*, 1771–1780. [CrossRef]
11. Xue, G.; Gelius, L.-J.; Sakyi, P.; Zhou, N.; Chen, W.; Su, B.; Li, H.; Zhong, H.; Su, Y. Discovery of a hidden BIF deposit in Anhui province, China by integrated geological and geophysical investigations. *Ore Geol. Rev.* **2014**, *63*, 470–477. [CrossRef]
12. Fu, J.; Jia, S.; Wang, E. Combined magnetic, transient electromagnetic, and magnetotelluric methods to detect a BIF-type concealed iron ore body: A case study in gongchangling iron ore concentration area, southern liaoning province, China. *Minerals* **2020**, *10*, 1044. [CrossRef]
13. Fu, G.; Yan, J.; Chen, H.; Meng, G.; Yu, H.; Fan, L.; Luo, L.; Tao, X. 3D inversion of audio-magnetotelluric data for mineral exploration: A case study of Layikeleke buried porphyry copper deposit, Xinjiang, China. *Appl. Geophys.* **2020**, *17*, 576–588.
14. Meng, G.; Lv, Q.; Yan, J.; Deng, Z.; Qi, G.; Xue, R. The research and application of explorational technology of "Penetrating" to geology and mineral investigation in overburden area. *Acta Geosci. Sin* **2019**, *40*, 637–650.
15. Yan, J.; Meng, G.; Lv, Q.; Deng, Z.; Qi, G.; Tang, H.; Zhao, J.; Xue, R.; Wang, X. Prediction and location of concealed deposits in desertgobi coverage areas using integrated geophysics: An example of the Layikeleke copper polymetallic deposit in Xinjiang, Northwest China. *Chin. J. Geophys.* **2021**, *64*, 4117–4133.
16. Wang, L.; Zhang, S.; Fang, Y.; Tang, L. Integrated Exploration Model for Concealed Ore Deposit: A Case Study from Shuitou Fluorite Deposit, Inner Mongolia, North China. *J. Earth Sci.* **2021**, *32*, 370–389. [CrossRef]
17. Fang, Y.; Zhang, S.; Zou, H.; Zhang, P.; Zeng, Z.; Gao, F. Comprehensive exploration method for fluorite deposits in grasslands covered area: A case study of the Saiboluogoumen fluorite deposit in Linxi, Inner Mongolia. *J. Chengdu Univ. Technol. (Sci. Technol. Ed.)* **2014**, *41*, 94–101.
18. Zou, H.; Li, M.; Santosh, M.; Zheng, D.; Cao, H.; Jiang, X.; Chen, H.; Li, Z. Fault-controlled carbonate-hosted barite-fluorite mineral systems: The Shuanghe deposit, Yangtze Block, South China. *Gondwana Res.* **2022**, *101*, 26–43. [CrossRef]
19. Wang, B.; Yang, X.; Li, S.; Teng, C.; Yang, X.; Huang, F.; Zhang, X.; Cao, J.; Zhou, Y.; Zhang, H. Geochronology, geochemistry, and tectonic implications of early Neoproterozoic granitic rocks from the eastern Beishan Orogenic Belt, southern Central Asian Orogenic Belt. *Precambrian Res.* **2021**, *352*, 106016. [CrossRef]
20. Wang, Y.; Zhang, F.; Li, B.; Xue, C.; Liu, J.; Zhao, Y.; Zhang, W. Geology and genesis of the Cihai mafic intrusions in Beishan Terrane, Xinjiang, Northwest China: Implication for iron mineralization and tectonic setting. *Ore Geol. Rev.* **2020**, *121*, 103573. [CrossRef]
21. Ruan, B.; Wei, W.; Yu, Y.; Lv, X. Geology, geochronology, mineral chemistry and geochemistry of the Hongnieshan mafic–ultramafic complex in the Beishan area, southern Central Asian orogenic Belt, NW China: Implications for petrogenesis and regional Ni mineralization. *Ore Geol. Rev.* **2021**, *139*, 104423. [CrossRef]
22. Yang, J.; Wang, L.; Wang, X.; Xie, X.; Jiang, A. SHRIMP zircon U-Pb dating of the Xiaohongshan vanadium-titanium magnetite deposit, Ejin Banner, Beishan, Inner Mongolia, and its geological implications. *Geol. Bull. China* **2015**, *34*, 1699–1705.
23. Zhi, J.; Lei, R.; Chen, B.; Muhtar, M.; Feng, Z.; Zhang, K.; Cai, Y.; Wu, C. Zircon genesis and geochronology for the Zhangbaoshan super-large rubidium deposit in the Eastern Tianshan, NW China: Implication to magmatic-hydrothermal evolution and mineralization processes. *Front. Earth Sci.* **2021**, *9*, 682720. [CrossRef]
24. Zhang, Z.; Feng, X.; Zhao, G.; Tan, W. The Metallogenic Rule of Mesozoic Hypogene Deposits in the Eastern Tianshan-Beishan Area. *Northwest Geol.* **2022**, *55*, 280–299.
25. Huo, M.; Lu, K.; Si, X.; Dai, S.; Lu, M.; Xue, Y.; Yao, L.; Liu, D.; Xie, G. Geological structural features and copper-gold prospecting direction of the northern Hutoushan in the eastern section of the Beishan Orogenic Belt, Inner Mongolia. *Miner. Explor.* **2020**, *11*, 2117–2125.
26. Pei, Q.; Zhang, S.; Hayashi, K.I.; Wang, L.; Cao, H.; Zhao, Y.; Hu, X.; Song, K.; Chao, W. Nature and genesis of the Xiaobeigou fluorite deposit, Inner Mongolia, northeast China: Evidence from fluid inclusions and stable isotopes. *Resour. Geol.* **2019**, *69*, 148–166. [CrossRef]
27. Li, K.; Shang, P.; Han, Z.; Wang, C.; Jiao, S.; Wang, G.; Yuan, C.; Zhang, Q.; Liu, Z.; Yan, X. Comprehensive Information Prospecting Method for Fluorite Deposits in Shaowu Area, Northern Fujian Province. *Acta Geosci. Sin.* **2022**, *43*, 404–410.

28. Li, K.; Chen, X.; Shang, P.; Yu, X.; Han, Z.; Wei, F.; Liu, Z. Application of the prospecting method of geophysical and geochemical integrated information in the exploration of fluorite deposits. *Geol. Surv. China* **2019**, *6*, 98–104.
29. Gao, F.; Zhang, S.; Zou, H.; Cao, H.; Wang, G.; Zeng, Z. The application of the field gamma ray spectrometry to the prospecting for fluorite deposits in LinXi area of Inner Monggolia. *Geophys. Geochem. Explor.* **2013**, *37*, 206–211.
30. Zou, H.; Zhang, S.; Fang, Y.; Chen, A.; Gao, F.; Tang, L.; Deng, M. The radioactive elements content of natural fluorite and its influence. *Geophys. Geochem. Explor.* **2014**, *38*, 478–484. [CrossRef]
31. Han, B.; Shang, P.; Gao, Y.; Jiao, S.; Yao, C.; Zou, H.; Li, M.; Wang, L.; Zheng, H. Fluorite deposits in China: Geological features, metallogenic regularity, and research progress. *China Geol.* **2020**, *3*, 473–489.
32. Orešković, J.; Milošević, A.; Kolar, S.; Borojević Šoštarić, S. The Žune Ba-F epithermal deposit: Geophysical characterization and exploration perspective. *Geol. Croat.* **2023**, *76*, 47–56. [CrossRef]
33. Müller, D.; Kwan, K.; Groves, D.I. Geophysical implications for the exploration of concealed orogenic gold deposits: A case study in the Sandy Lake and Favourable Lake Archean greenstone belts, Superior Province, Ontario, Canada. *Ore Geol. Rev.* **2021**, *128*, 103892. [CrossRef]
34. Hronsky, J.M.; Groves, D.I. Science of targeting: Definition, strategies, targeting and performance measurement. *Aust. J. Earth Sci.* **2008**, *55*, 3–12. [CrossRef]
35. Adagunodo, T.; Sunmonu, L.; Adeniji, A. An overview of magnetic method in mineral exploration. *J. Glob. Ecol. Environ.* **2015**, *3*, 13–28.
36. Yu, W.; Wang, M.; Shang, P.; Huang, J.; Wang, Y. Application of high-accuracy ground magnetic survey to the exploration of fluorite deposits. *J. Geol.* **2019**, *3*, 428–433.
37. Sharma, S.; Biswas, A.; Baranwal, V. Very low-frequency electromagnetic method: A shallow subsurface investigation technique for geophysical applications. In *Recent Trends in Modelling of Environmental Contaminants*; Sengupta, D., Ed.; Springer: New Delhi, India, 2014; pp. 119–141.
38. Fang, Y.; Zhang, S.; Zou, H.; Wang, G.; Cao, H.; Li, H. The application of VLF and AMT to the exploration of fluorite deposit. *Geophys. Geochem. Explor.* **2013**, *37*, 800–803.
39. Zhang, Y.; Wang, L.; Gao, Y. Geological characteristics, types and development and utilization value of Shuitou fluorite deposit in Linxi, Inner Mongolia. *Geol. China* **2023**, *50*, 795–805.
40. Chen, W.; Hao, Q.; Chu, S.; Liu, J.; Liu, H.; Jiang, X. Application of very low frequency electromagnetic method to positioning of concealed metal deposits: An example of copper polymetallic ore occurrences in the southwest greater Hinggan mountains. *Geol. Explor.* **2017**, *53*, 528–532.
41. Xia, B.; Cao, H.; Pei, Q.; Yang, B.; Wu, Z.; Xu, T.; Han, S. Exploration of fluorite deposit by the combination of VLF-EM and EH4 on shallow-covered area. *J. Guilin Univ. Technol.* **2016**, *36*, 228–233.
42. Liu, P.; Li, Z.; Sang, J.; Zhao, L.; Tie, W. Radioactive characteristics and prospecting significance of fluorite deposits in Southern Song County of Western Henan Province. *Miner. Explor.* **2024**, *15*, 1015–1024. [CrossRef]
43. Ge, L.; Li, F. Review of in situ X-ray fluorescence analysis technology in China. *X-ray Spectrom.* **2020**, *49*, 458–470. [CrossRef]
44. Liu, C.; Sun, B.; Wei, L.; Li, H.; Meng, F.; Dong, F. Application of the integrated electrical method to exploration in the Zhaishang gold deposit, West Qinling. *Geol. Explor.* **2020**, *56*, 1226–1237.
45. Liu, C.; Li, H.; Sun, B.; Ye, G.; Hao, Z.; Xue, D.; Deng, A. Comprehensive geophysical study of Zhaishang Gold deposit, West Qinling. *Geol. Rev.* **2022**, *68*, 658–672. [CrossRef]
46. Pitiya, R.; Lu, M.; Chen, R.; Nong, G.; Chen, S.; Yao, H.; Shen, R.; Jiang, E. Audio Magnetotellurics Study of the Geoelectric Structure across the Zhugongtang Giant Lead–Zinc Deposit, NW Guizhou Province, China. *Minerals* **2022**, *12*, 1552. [CrossRef]
47. Egbert, G.D. Robust multiple-station magnetotelluric data processing. *Geophys. J. Int.* **1997**, *130*, 475–496. [CrossRef]
48. Rodi, W.; Mackie, R.L. Nonlinear conjugate gradients algorithm for 2-D magnetotelluric inversion. *Geophysics* **2001**, *66*, 174–187. [CrossRef]
49. Cai, J.; Chen, X. Refined techniques for data processing and two-dimensional inversion in magnetotelluric II: Which data polarization mode should be used in 2D inversion. *Chin. J. Geophys.* **2010**, *53*, 2703–2714.
50. Wang, G.; Fang, H.; Qiu, G.; Huang, J. The deep electrical structure across Anqing-Guichi ore concentration area. *Geol. China* **2019**, *46*, 795–806.
51. Egbert, G.D.; Kelbert, A. Computational recipes for electromagnetic inverse problems. *Geophys. J. Int.* **2012**, *189*, 251–267. [CrossRef]
52. Kelbert, A.; Meqbel, N.; Egbert, G.D.; Tandon, K. ModEM: A modular system for inversion of electromagnetic geophysical data. *Comput. Geosci.* **2014**, *66*, 40–53. [CrossRef]
53. Jiang, F.; Chen, X.; Unsworth, M.; Cui, T.; Sun, X.; Han, B. Comparison of 2D and 3D inversion results: A case study of an MT profile. In Proceedings of the 24th EM Induction Workshop, Helsingør, Denmark, 12–19 August 2018.
54. Singh, S.; Rani, A.; Mahajan, R.K. 226Ra, 232Th and 40K analysis in soil samples from some areas of Punjab and Himachal Pradesh, India using gamma ray spectrometry. *Radiat. Meas.* **2005**, *39*, 431–439. [CrossRef]
55. Yang, Y.; Zhang, G.; Yao, C.; Deng, Z.; Ren, Z.; Li, C. Application of Induced Polarization Method in Mineral Resource Exploration. *Sustainability* **2023**, *15*, 3840. [CrossRef]

56. Zhang, C.; Li, S.; Bai, D.; Cheng, H.; Cao, J.; Zhang, A.; Sun, J.; Zhao, H. Application of the time—Domain IP method to the exploration of concealed fluorite deposits:A case study of the Tielu fluorite deposit in Fangcheng County, Henan Province. *Geol. Explor.* **2022**, *58*, 369–380.
57. Wang, J.; Shang, P.; Xiong, X.; Yang, H.-Y.; Tang, Y. Metallogenic regularities of fluorite deposits in China. *Geol. China* **2015**, *42*, 18–32. [CrossRef]

minerals

MDPI

Article

Deep Geological Structure Analysis of the Dongyang Area, Fujian, China: Insights from Integrated Gravity and Magnetic Data

Zhenyu Zhang [1,2,3], Yongbo Li [1,2,3,4,*], Liang Chen [5], Qiang Zhang [1,2,3] and Yue Sun [1,2,3]

[1] Institute of Geophysical and Geochemical Exploration, Chinese Academy of Geological Sciences, Langfang 065000, China; zzy8324@126.com (Z.Z.)
[2] National Center of Geological Exploration Technology, Langfang 065000, China
[3] Key Laboratory of Geophysical Electromagnetic Probing Technologies, Ministry of Natural Resources, Langfang 065000, China
[4] Chinese Academy of Geological Sciences, Beijing 100037, China
[5] Center for Geophysical Survey, China Geological Survey, Langfang 065000, China
* Correspondence: lyongbo@mail.cgs.gov.cn

Abstract: To explore the deep geological structure of the Dongyang area in Fujian, China, gravity data from the area and its surroundings were collected and processed. Additionally, a high-precision magnetic survey was conducted in the Zhongxian region of this area, with subsequent analysis of the magnetic anomalies. Through the integration of regional geological data, a comprehensive analysis was carried out on the characteristics of gravity–magnetic anomalies and deep geological structures in the Dongyang area. The study indicates that the primary portion of the Dongyang area lies southwest of the expansive circular volcanic structure spanning Dehua to Yongtai. Two significant residual gravity anomalies were identified within the region, interpreted as the Xiaoban-Shuangqishan and Dongyang-Lingtouping residual gravity-positive anomalies. In the Zhongxian region, the magnetic field exhibits complexity with notable amplitude variations. Positive anomalies predominate in the western and northern sectors, while localized positive anomalies are prominent in the eastern region. The central area portrays a circular and disordered mix of positive and negative anomalies. Particularly distinctive are the band-shaped and fan-shaped negative anomalies curving from northeast to southeast through the central region. Various positive and negative anomalies of varying strengths, gradients, and orientations overlay both positive and negative magnetic backgrounds in specific locales. Moreover, the Dongyang area showcases well-developed fault structures, primarily oriented in northeast and northwest directions. Leveraging the regional magnetic attributes in conjunction with regional geological data, 39 faults were deduced in the Zhongxian region of the Dongyang area, delineating three promising mineralization zones.

Keywords: Dongyang area; magnetotelluric belt; magnetic exploration; deep geological structures; gold deposits

1. Introduction

Magnetic and gravity exploration is a geophysical method based on the density and magnetic differences between various rocks, minerals, or other geological targets. It is used to study the deep-seated structures, regional tectonics, crystalline basement, and regional major faults, and to delineate metal and non-metal mineralization belts [1–4]. When geological conditions are favorable, magnetic and gravity exploration is one of the most effective combinations of geophysical methods for direct mineral exploration [5–7].

Located in southeastern China, Fujian Province has been shaped by numerous orogenic movements, with the Yanshan movement exerting a significant influence on regional tectonics, plate evolution, and mineralization processes [8,9]. The province is geologically segmented into three distinct units: the northwest, southwest, and eastern regions, demarcated by the NE-trending Nanping-Ninghua fault zone and the NNE-trending Zhenghe-Dapu structural magmatic belt [10,11]. Within this intricate geological landscape lies the Dongyang area, positioned at the heart of eastern Fujian. Here, it intersects the southern segment of the Zhouning-Hua'an fault uplift belt and the NE-trending Yangmei-Donghua fault, along with the NW-trending Ancun-Gujiakou fault to the south. Characterized by widespread volcanic rock formations, the region boasts a rich abundance of gold deposits, notably epitomized by the renowned "Golden Triangle" of Fujian Province in the Yuxi-Dehua-Yongtai gold ore prospecting area [12,13].

The Dongyang gold deposit is part of the circum-Pacific metallogenic belt, a legacy of the Yanshan mid-term volcanic activities and the subduction of oceanic plates beneath the Eurasian plate [14]. Evolving through multiple tectonic phases post the Jinning movement, the Dongyang metallogenic belt delineates a structural schema marked by profound or major faults, complemented by an intricate network of secondary faults. These faults converge in a belt-like configuration, predominantly manifesting as the NE-NNE and NW-NNW fault zones within the area [15]. The mineralization of the Dongyang gold deposit is intimately tied to the Late Jurassic–Early Cretaceous volcanic events. Spatially constrained along the margins of large volcanic structures, ore bodies predominantly align along major faults, minor intrusive bodies, and basins, collectively exhibiting a prevailing NW-oriented belt-like distribution [16].

Dongyang's geological landscape is characterized by an extensive presence of volcanic rocks, intense magmatic activities, and a diverse array of mineral deposits [17,18]. Presently, the region boasts the discovery of over 20 mineral deposits, including notable sites such as the Dehua Shuangqishan gold deposit, Qiucun gold deposit, and Youxi Xiaoban gold deposit, underscoring a vast exploration potential within the region's deep subsurface [19,20]. Regional geophysical exploration and the pursuit of deep-seated mineralization in the Dongyang area have emerged as focal points within the realm of cutting-edge national mineral exploration research [21].

The Dongyang area is located at the intersection of multiple tectonic zones, where fault structures mainly control mineralization belts. The area is heavily dissected by the terrain, with a relatively low level of geophysical exploration and limited understanding of the deep-seated geological structures. This study conducted magnetic exploration in the Zhongxian region of Dongyang, Fujian, combining regional gravity anomalies to investigate the deep-seated fault structures in the central Dongyang region. The findings provide a basis for exploring deep-seated mineralization in the area.

2. Geologic Setting

The Dongyang area is situated within the Shouning-Hua'an uplift belt and the Fuding-Pinghe fault depression belt, both classified as fifth-order tectonic units [22]. Predominantly positioned in the middle segment of the Shouning-Hua'an uplift belt, the area's structural composition converges at the intersection of the Zhouning-Hua'an uplift belt with the Pingnan-Meilin fault depression belt and the Fuding-Yunxiao fault depression belt. Bounded by the Zhenghe-Dapu fault zone and the Fu'an-Nanjing Beidong fault zone, the Dongyang area is traversed by the NNE-trending Pucheng-Youxi fault zone in its western expanse [23]. Throughout geological evolution since the Paleoproterozoic era, the region has undergone successive episodes of tectonic, magmatic, and mineralization transformations, fostering intricate relationships between tectonic settings and mineralization dynamics across various geological stages. These conditions have been instrumental in nurturing the abundant gold mineral resources within the Dongyang area. Figure 1 shows the geological mineral map of the Zhongxian region in the Dongyang area.

2.1. Stratigraphy

The area exposes five stratigraphic formations, including the Neoproterozoic formation; the Sinian formation; the Late Triassic–Middle Jurassic formation; the Late Jurassic–Cretaceous continental deposits; and the Quaternary loose accumulation layer. Mineral deposits in the region are predominantly governed by the secondary NW-trending structures of the Da Yunshan giant volcanic ring structure. Ore bodies are delineated by NW-trending and NE-trending faults, manifesting in vein-like and lens-like structures. Noteworthy ore-bearing strata in the region include the Neoproterozoic Dalingshan Formation, Late Jurassic Nanyuan Formation, and Late Jurassic Changlin Formation, all closely associated with gold mineralization [24,25].

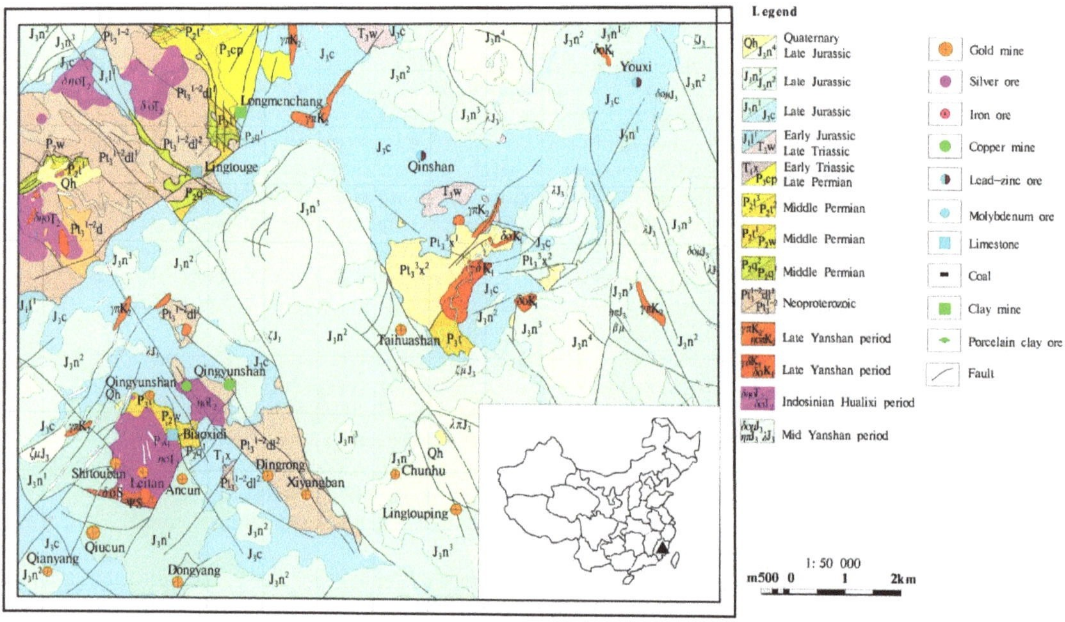

Figure 1. Geological mineral map of the Zhongxian region in the Dongyang area (modified from the 1:50,000 geological mineral map of the Dongyang area, Fujian).

2.2. Structure

The Dongyang area has been a focal point of intense tectonic magmatic activity across various geological periods, including the Caledonian, Hercynian, Indosinian, Early Yanshan, Middle Yanshan, Late Yanshan, and Xishan epochs. Notably, the Early and Late Yanshan activities were characterized by extensive multi-stage and multi-phase volcanic events. Intrusive rocks predominantly manifest as plutons and stocks, with occasional occurrences as plutons or dykes [26]. The prevalent rock types comprise granodiorite, diorite granite, and granodiorite, typically exhibiting a belt-like distribution trending NE. The region's magmatic sources exhibit a diverse origin, sourced from basic rocks and granites in the upper mantle, with a mixed mantle–crustal provenance, as well as magmas resulting from crustal remelting processes [27,28].

The regional structural framework is significantly influenced by the Mingxi southwest depression, the Mindong volcanic fault depression belt, and the three NNE-trending Zhenghe-Dapu, Fu'an-Nanjing, and Pucheng-Youxi fault zones. The Haixi period was characterized by frequent oscillatory activities, often giving rise to detachment structures along major lithological boundaries. From Indosinian to Yanshanian orogenies, tectonic–magmatic interactions engendered folding, thrust structures, reconfiguration of pre-existing stratigraphic and structural surfaces, and modification of earlier mineralized bodies, culminating in a spectrum of polymetallic deposits encompassing gold, silver, copper, lead, zinc, manganese, iron, and other elements. Furthermore, a series of brittle fault zones characterized by north-northeast trends were established [29–31].

3. Methodology

Due to the fact the deep-seated mineral deposits in the Dongyang area are primarily controlled by fault structures, this study conducted a comprehensive analysis of 1:200,000 regional gravity data obtained from the Dongyang area and its periphery. Furthermore, the ore-controlling fault structures are typically located in the transitional zone of intense variations in positive and negative magnetic fields, and a high-precision ground magnetic survey at a scale of 1:50,000 in the Zhongxian region of the Dongyang area was also carried out.

3.1. Regional Gravity

The gravity data underwent processing utilizing the interpolation-cutting method. The Bouguer gravity anomaly map (Figure 2a) delineates the structural characteristics of the Dehua-Yongtai giant ring-shaped volcanic structure. The outer ring exhibits several subtle high-gravity anomalies, while the inner ring is distinguished by a prominent large-scale low-gravity anomaly encircled by the outer ring. The demarcation between the inner and outer rings displays a certain degree of indistinctness on the Bouguer gravity anomaly map, possibly attributed to minor variations in the physical properties of corresponding geological formations.

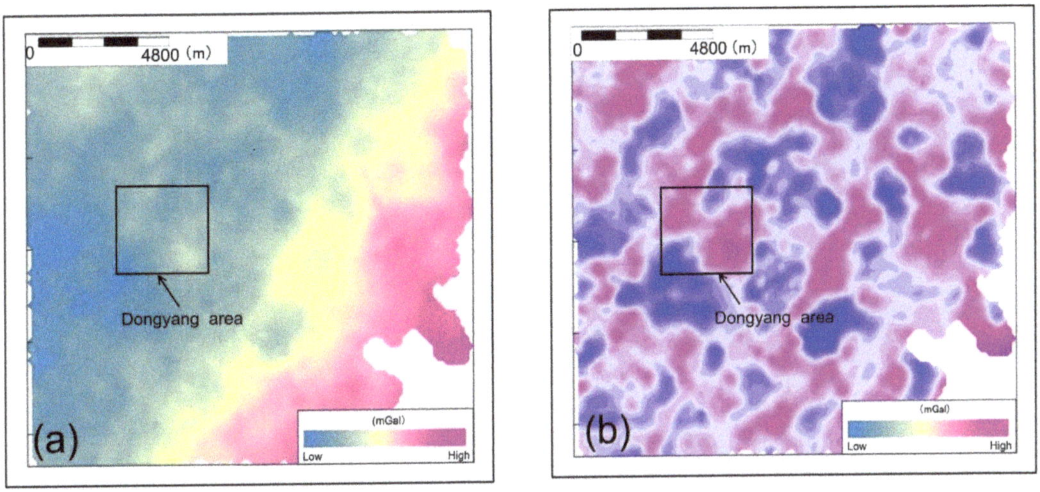

Figure 2. (**a**) The Bouguer gravity and (**b**) the residual gravity anomaly maps of the Dongyang area and its periphery. The black rectangular area denotes the Dongyang area.

From the Bouguer gravity anomaly map and the residual gravity anomaly map (Figure 2), it was discerned that the main body of the Dongyang area is situated in the southwestern inner ring of a giant circular structure. Within this area, two significant residual gravity anomalies are conspicuously distributed. Given the regional geology, it is postulated that these anomalies are correlated with the Xiaoban-Shuangqishan gravity anomaly and the Dongyang-Lingtouping gravity anomaly [32,33].

To investigate the regional deep structures of the Dongyang area and its periphery, this study conducted upward continuations of gravity anomalies to various heights. Subsequently, vertical derivatives of theta maps were calculated to detect the edge [34], enabling multi-scale edge detection. By utilizing boundary information derived from different depths of the geological structure, the distribution of deep-seated fault structures was examined (refer to Figure 3).

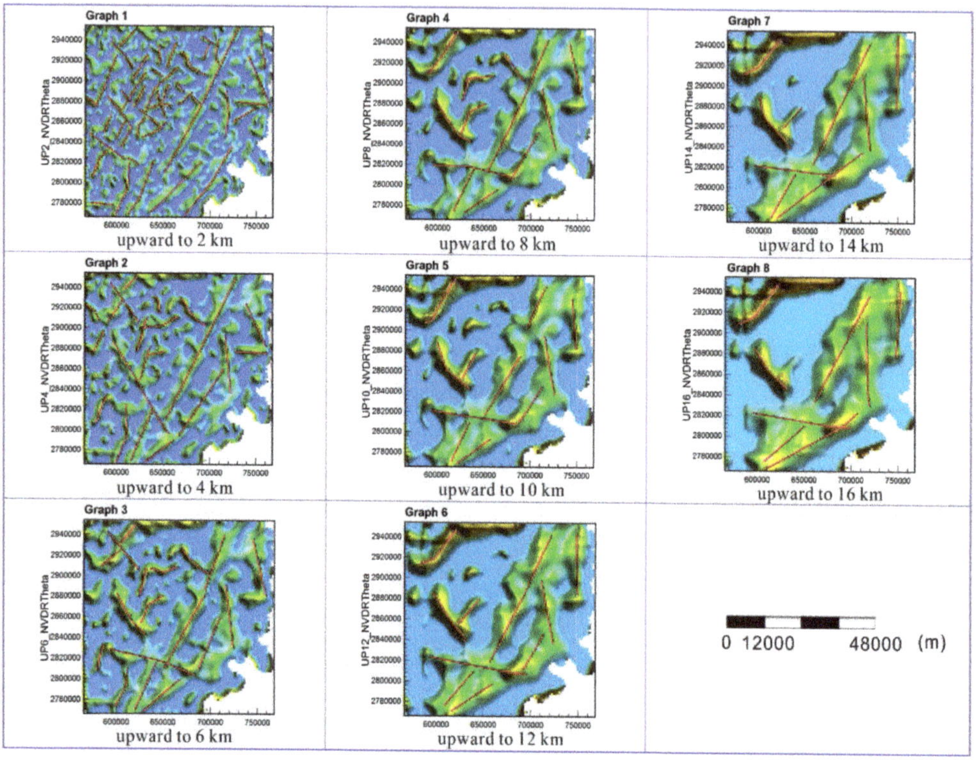

Figure 3. The multi-scale boundary identification and inference of linear structure from the vertical derivative maps of Theta upward by different heights. The Theta values come from the gravity data, and the red lines represent inferred faults.

3.2. Regional Magnetic Field

In this study, the total magnetic intensity (TMI) after the international geomagnetic reference field (IGRF) was gridded using the Kriging interpolation method with a grid size of 250 m $\times$ 100 m. The reduction to the pole (RTP) was transformed using the frequency domain with the inclination and declination of 39.24° and −4.18°, respectively [35–42]. Furthermore, first-order derivatives of the RTP data were enhanced in four horizontal directions (0°, 45°, 90°, 135°), and upward continuation was also carried out [43–45].

The magnetic anomaly maps of the Zhongxian region (refer to Figure 4) distinctly revealed the presence of a circular volcanic structure in the central region, along with showcasing the distribution characteristics of regional fault structures. Examination of the horizontal directional derivative maps (see Figure 5) in the region emphasized the distribution patterns of various fault structures in different orientations.

Magnetic susceptibility inversion is a data processing technique employed to deduce the physical characteristics and spatial configuration of subsurface materials by analyzing magnetic parameters acquired at the Earth's surface [46,47]. In this study, a three-dimensional (3D) magnetic susceptibility inversion was executed on magnetic survey data utilizing the MAG3D 4.0 software [48], developed by Columbia University. The inversion was conducted utilizing the original 1:50,000 magnetic RTP data of the volcanic structure in the Zhongxian region without constraint. The software automatically generated a mesh file comprising 84 grids both in the EW (X-axis) and NS (Y-axis) directions, spaced at 100-meter intervals, and 70 vertical grids in the Z direction with increasing spacings in a 50 m $\times$ 4 grid, 100 m $\times$ 10 grid, 200 m $\times$ 15 grid, and 400 m $\times$ 5 grid. The model size was

8400 m × 8400 m × 8000 m (X × Y × Z). Default values for the initial and reference models were zero, with a weighting factor of 2 for the depth weighting function.

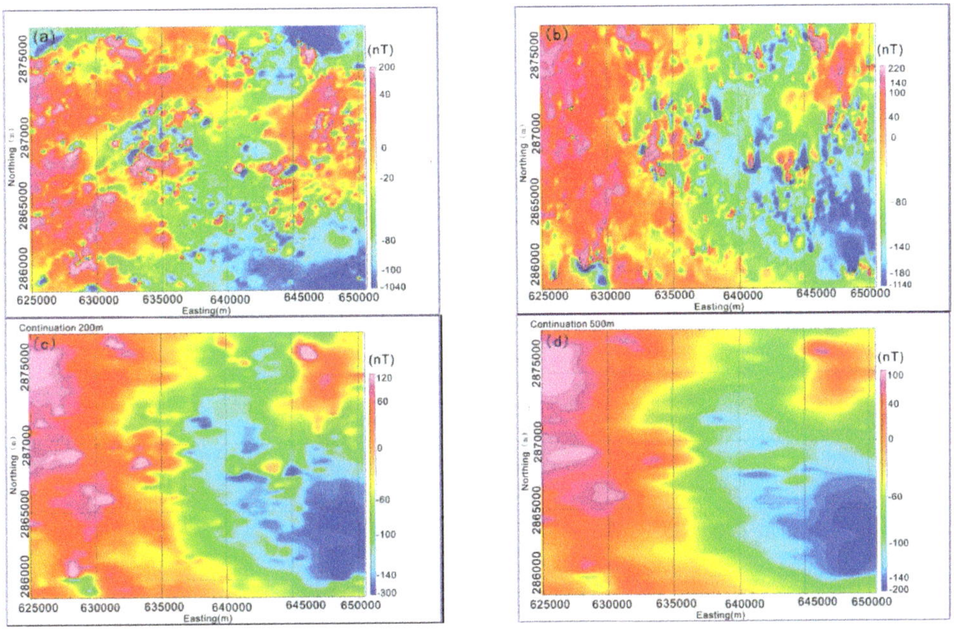

Figure 4. The magnetic anomaly maps of (**a**) total magnetic intensity (TMI) after international geomagnetic reference field (IGRF), (**b**) reduction to the pole (RTP), upward continuation to (**c**) 200 m and (**d**) 500 m in the Zhongxian region of the Dongyang area.

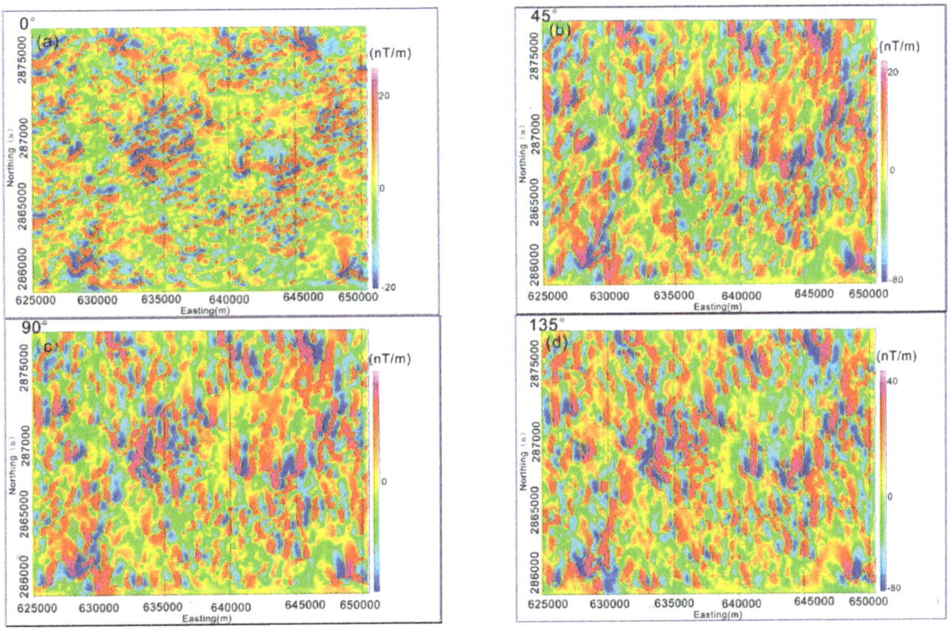

Figure 5. The horizontal directional derivative maps of (**a**) 0°, (**b**) 45°, (**c**) 90°, and (**d**) 135°.

The magnetic susceptibility inversion model is depicted in Figure 6, while the inversion outcomes are illustrated in Figure 7. The inversion results exhibit a good correlation between the distribution of magnetic bodies (intermediate basic volcanic rocks) linked to the subsurface volcanic structure and the TMI after IGRF (Figure 4a). The presence of magnetic bodies (intermediate basic volcanic rocks) within the underground volcanic structure of the study region suggests a promising potential for deep-seated mineral exploration [49,50].

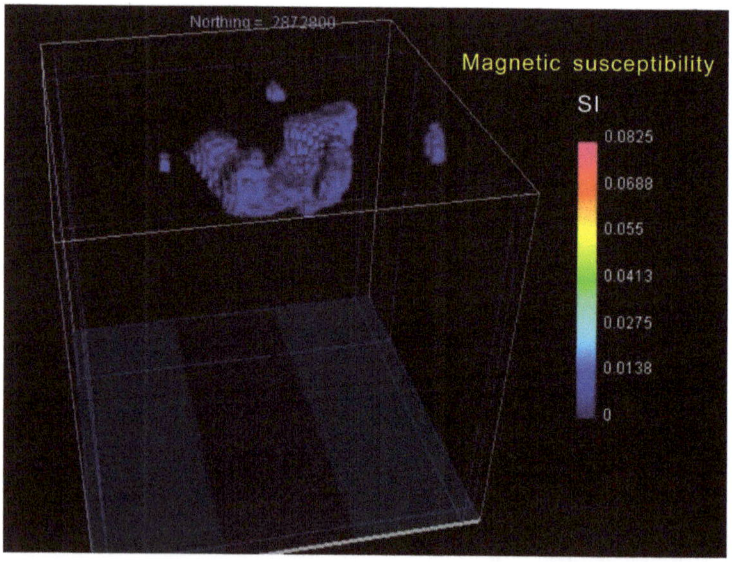

Figure 6. The 3D magnetic susceptibility inversion model. The magnetic susceptibility of the blue isosurface is 0.01.

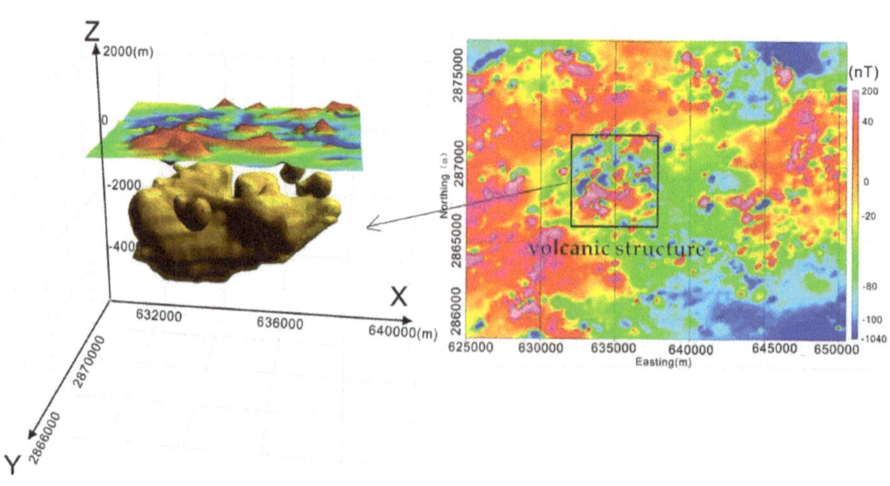

Figure 7. The magnetic bodies of the volcanic structure in the Zhongxian region by 3D magnetic susceptibility inversion. The right figure is the TMI map after IGRF of the Zhongxian region, and the rectangular area denotes the volcanic structure in the Zhongxian region.

4. Results and Discussion

4.1. Regional Gravity Field Analysis

From the theta vertical derivative maps (Figure 8), it is apparent that the Dongyang area has two sets of NE-trending and NW-trending faults. Furthermore, a radial fault system is discernible in the NE corner of the area. Both the NE-trending and NW-trending faults exhibit distinct features in the theta vertical derivative map, showcasing an upward continuation of 2 km (Figure 8a). As the upward height increases, each of the two NE-trending and two NW-trending faults converges into a single fault in their respective directions. The plane projections of the NW-trending faults shift southwestward, while the NE-trending faults move southeastward, delineating the fault trends. Concurrently, the radial fault system gradually diminishes (Figure 8b), indicating constrained extension space for radial faults, a characteristic trait of volcanic structures. This observation implies the potential presence of a substantial concealed volcanic structure in the Dongyang area.

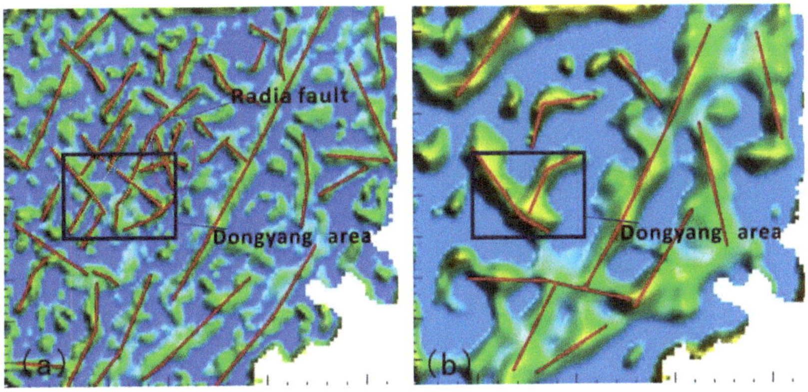

Figure 8. The inferred structures from the theta vertical derivative maps with upward continuations of (**a**) 2.0 km and (**b**) 8.0 km. The red lines represent inferred faults, the black rectangular area denotes the Dongyang area, and the purple dashed lines represent two sets of NE-trending and NW-trending faults.

As the extension height escalates, the linear structures identified by the theta vertical derivative progressively diminish, emphasizing the enhanced representation of deep major fault information and the disappearance of shallow fault details. The multi-scale edge detection of the gravity field effectively unveils the spatial extension characteristics of the primary faults within the Dongyang area and its periphery.

4.2. Regional Magnetic Field Analysis

Based on the TMI after IGRF of the magnetic survey in the Zhongxian region (Figure 4a), the regional magnetic field displays a high degree of complexity, characterized by significant variations in amplitude. Predominantly positive anomalies are observed in the western and northern sectors, while localized positive anomalies are prevalent in the eastern region. A circular pattern of chaotic positive and negative anomalies is discernible in the central area, featuring a distinctive fan-shaped negative anomaly extending from the northeast through the central region to the southeast. Across both positive and negative magnetic field backgrounds, anomalies of varying strength, steepness, and orientations overlap.

The unique characteristics of the regional magnetic field allow for the division of the Zhongxian region into three distinct magnetic zones: Longmen-Donghua-Yangmei (Zone I), Shangzhuang-Huashan-Chunhu-Gaiyang (Zone II), and Jihua-Chiling (Zone III). Zone I is further subdivided into two subzones, as illustrated in Figure 9.

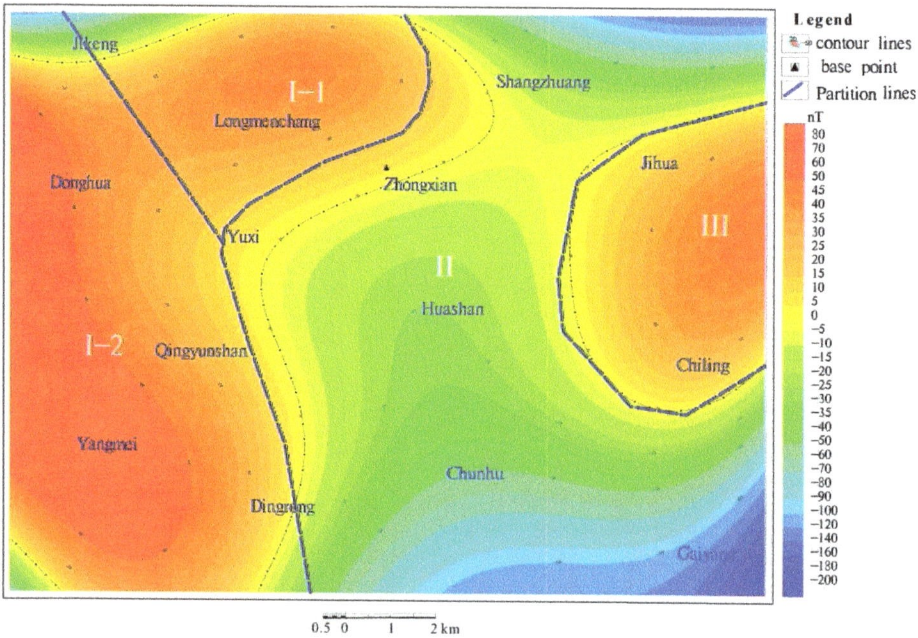

Figure 9. The magnetic field zoning map of the Zhongxian region. The blue dashed lines represent the magnetic field zones with their number.

(1) Zone I: Longmen-Donghua-Yangmei

Located in the western and northern parts of the Zhongxian region, Longmenchang-Donghua-Yangmei (Zone I) exhibits a large magnetic control area spanning 18 km in length and 12 km in width, with a magnetic anomaly amplitude ranging from −100 to 90 nT. This zone is characterized by the superposition of two positive anomalies and an NW-trending reduced anomaly band. The outer boundaries are delineated by different magnetic zones and gradient bands, with the anomalies being more pronounced on the western side compared to the northern side.

The area is further divided into subzones I-1 and I-2 by a reduced anomaly band running along the Jikeng-Yuxi line. The I-1 subzone, elliptical in shape, trends northeastward, while the I-2 subzone, a strip widening from west to east, trends northwestward with a less distinct northward reduced anomaly band in the middle. The geological background of this magnetic zone is complex, comprising the Late Jurassic Nanyuan Formation and Changlin Formation, Early Jurassic Lishan Formation volcanic rocks, Permian Cuipingshan Formation, Tongziyan Formation, Wenbishan and Qixia Formations, and Nanhua-Qingbaikou Period Mamianshan Group Dalingshan Formation metamorphic rocks, as well as intrusive medium- to fine-grained quartz diorite and quartz monzonite of the Early Caledonian period. Based on geological data, the main uplift in the I-1 subzone is inferred to be associated with Early Caledonian quartz diorite uplift, while the main feature in the I-2 subzone is linked to Early Caledonian quartz monzonite basement. The NW- and NE-trending reduced anomaly bands connecting the outer gradient bands of the magnetic zone are interpreted as representing deeply incised magnetic basement fault structures.

(2) Zone II: Shangzhuang-Huashan-Chunhu-Gaiyang

Situated in the central part of the Zhongxian region, Shangzhuang-Huashan-Chunhu-Gaiyang (Zone II) extends from Shangzhuang in the northeast, through Huashan in the central region, to Gaiyang in the southeast. The magnetic zone forms a "C" shape, spanning 18 km in length and 10 km in width, with a northward and northwestward distribution.

The two directions converge near Huashan, with the magnetic field transitioning to a northward extension near Gaiyang. The zone predominantly exhibits negative magnetic characteristics, with a gradual variation in the central area and significant gradient changes on the northern and southern sides. The amplitude varies with higher values in the central region and lower values at both ends, reducing from 0 to −200 nT.

The area primarily exposes the second, third, and fourth sections of the Nanyuan Formation and Changlin Formation, with the western Xixi Formation metamorphic rocks cropping out in the central region. Intrusions of granite porphyry, granodiorite, granite diorite, and quartz diorite are observed in Shangzhuang and Heshun. Based on geological data, the main features of the magnetic zone are interpreted as volcanic rock cover and the non-magnetic or weakly magnetic Changlin Formation. The relatively high anomaly amplitude in the central region is indicative of uplifted acidic-intermediate-neutral rock bodies, while the northward and northwestward magnetic trends are attributed to different fault structures.

(3) Zone III: Jihua-Chiling

Situated in the eastern part of the Zhongxian region, the Jihua-Chiling Zone exhibits a more compact magnetic range, measuring 8.6 km in length and 6.8 km in width. With a semi-elliptical shape, this zone features anomalies trending NE and predominantly displaying positive magnetic characteristics, with amplitudes ranging from 0 to 50 nT. Surface geology in this zone primarily corresponds to the second and third sections of the Nanyuan Formation volcanic rocks, with localized intrusions of late Yanshan granodiorite and exposures of late Jurassic rhyolite. The magnetic field variations are attributed to lithological changes, volcanic rock formation conditions, as well as tectonic and magmatic activities, indicating a prevalence of uplifted granodiorite basement in the magnetic field interpretation.

This detailed zoning and analysis of the magnetic field within the Zhongxian region offer valuable insights into the geological structures and potential mineral resources, significantly enhancing our comprehension and exploration of the region's subsurface characteristics.

4.3. Interpretation of Fault Structures

The fault structures within the magnetic field exhibit a diverse range of characteristics, manifesting mainly in eight distinct types: boundaries between different magnetic field zones, magnetic anomaly gradient zones, bead-like magnetic anomaly zones, linear anomaly zones, magnetic anomaly abrupt zones, anomalous displacement zones, staggered anomaly zones, and radial groups of anomaly zones. Through an analysis of the magnetic field characteristics of TMI after IGRF, RTP, and upward continuations to 200 m and 500 m in Figure 4, in conjunction with regional gravity field features, a total of 39 fault structures have been inferred within the Zhongxian region. These fault structures predominantly trend in the NE, NNE, NW, and NNW directions, with some also exhibiting near north–south orientations, as shown in Figure 10.

Within the Zhongxian region, magnetic surveys have identified a total of 19 fault structures trending NE and NNE, including 7 newly recognized faults. NE-trending faults often intersect obliquely with the regional magnetic field, leading to discontinuous magnetic field patterns that frequently manifest as boundaries between different magnetic field zones, linear gradient zones, or bead-like anomalies. Major NE-trending faults such as F8 and F13 exhibit regional characteristics, acting as boundaries within the magnetic field zones. These faults, especially near intersections with NW-trending faults, significantly influence the distribution of gold, lead, zinc, and other polymetallic deposits in the region, with the Qiu Village gold deposit primarily associated with the NNE-trending fault F27.

Similarly, within the Zhongxian region, magnetic surveys have identified 18 fault structures trending NW and NNW, including 8 newly identified faults. NW-trending faults commonly function as boundaries between different magnetic field zones, linear gradient zones, and low-magnetic-value zones, causing discontinuities along the magnetic field. Major NW-trending fault zones, such as F4 and F37, exhibit regional structural significance, with F4 acting as a boundary within the magnetic field zones and F37 closely aligning with geological structures, representing a deeply incised magnetic basement fault structure. These NW-trending fault zones play crucial roles as conduits for the migration of ore-forming elements, influencing the distribution of gold, lead, zinc, and other polymetallic deposits in the region, holding regional ore-controlling importance. Notably, the Xiqian copper–gold deposit and Chunhu gold deposit are believed to be controlled by the NW-trending F29 and F32 fault structures, respectively.

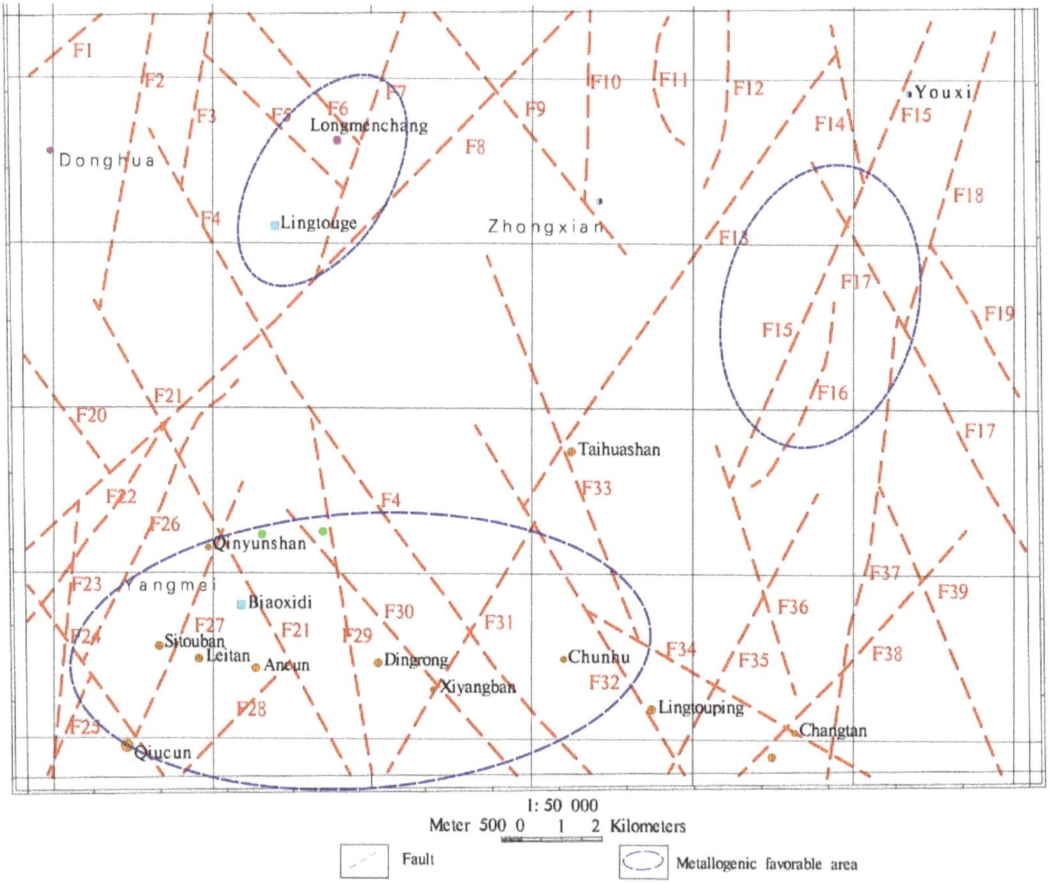

Figure 10. Fault structure interpretation of the Zhongxian region. The red dashed lines represent inferred fault structures, and the areas enclosed by blue dashed circles denote metallogenically favorable areas.

Furthermore, magnetic surveys in the area have identified two nearly NS-trending faults, including one newly recognized fault. NS-trending faults often exhibit characteristics such as bead-like magnetic anomalies and are typically located close to volcanic vents, with F12 forming part of a circular volcanic structure.

4.4. Prospective Mineral Exploration Area

Based on the magnetic and gravity characteristics, in conjunction with the interpreted results and consideration of the known mineral deposits in the Zhongxian region, three promising exploration blocks have been identified with the potential for discovering polymetallic deposits in the region.

(1) Longmenchang Prospective Exploration Area

The Longmenchang area showcases multiple sets of NE- and NW-trending fault structures, including two inferred NE-trending and two NW-trending fault structures. Surface exposures along these structures reveal widespread occurrences of limonite and silicified alteration. At the intersections of these structures, lead–zinc mineralization, limestone, and clay small-scale deposits have been observed. The favorable exploration conditions in this area suggest significant potential for mineralization. Exploration efforts should prioritize investigating relevant mineral deposits near fault and lithological contact zones.

(2) Qingyunshan-Qiucun-Chunhu Prospective Area

Within this area, 14 faults have been identified, consisting of seven NE-trending and seven NW-trending structures. Surface occurrences of silicification, pyritization, and limonitization alterations are prevalent. At the intersections of these structures, four gold mineralization points, six small-scale gold deposits, and one medium-sized gold deposit have been discovered. Given these findings, the exploration conditions in this area are considered favorable. Exploration activities should emphasize the exploration of concealed gold deposits by integrating alteration zone factors at fault contact zones and rock boundaries.

(3) Youxijihua-Yongtai Zhufeng Prospective Area

In this region, fault structures predominantly trend in the NE and ENE directions. One NE-trending and one NW-trending structure have been inferred, with the NE-trending structure identified as a significant ore-controlling structure. Intensive surface alteration is evident, characterized by silicification, pyritization, and limonitization alterations. The overall assessment indicates favorable exploration conditions in this area. Prospecting endeavors should focus on exploring concealed gold mineralization near lithological contact zones and structural belts.

5. Conclusions

The Dongyang area is strategically located within the "Golden Triangle" gold deposit area of Fujian Province, China, characterized by exceptional metallogenic geological settings and promising mineralization potential. This study meticulously gathered and processed gravity data from the Dongyang area and its periphery. Additionally, a precise 1:50,000 magnetic survey was carried out in the Zhongxian region of the Dongyang area. The research involved comprehensive data processing and information extraction from the gravity and magnetic datasets, facilitating a thorough examination of the deep-seated geological structures in the Zhongxian region.

The findings reveal a well-developed network of deep-seated fault structures in the Zhongxian region, predominantly oriented in the northeast and northwest directions. A total of 39 inferred faults were delineated, encompassing 19 NE- and NNE-trending faults, 18 NW- and NNW-trending faults, and 2 nearly NS-trending faults. Through the integration of regional geological data, three prospective mineral exploration zones were identified.

This study offers initial insights into the deep-seated geological structural characteristics of the region, providing valuable geophysical data and a foundation for extensive mineral exploration endeavors in the Dongyang area. Furthermore, it holds significant implications for advancing the understanding of metallogenic theories and mineralization patterns within the "Golden Triangle" gold deposit region of Fujian, China.

This study only focuses on gravity and magnetic exploration, inferring the deep-seated fault structures in the Zhongxian region based on gravity and magnetic features. It provided a preliminary understanding of the deep-seated geological structures in the Zhongxian region. To obtain better understanding of the regional deep-seated geological structures, further integrated geophysical exploration methods such as magnetotelluric sounding and seismic surveys need to be conducted.

Author Contributions: Conceptualization, Z.Z. and Y.L.; methodology, Z.Z. and L.C.; software, Z.Z. and L.C.; investigation and validation, Z.Z., Y.L. and L.C.; writing—original draft preparation, Z.Z. and Y.L.; writing—review and editing, all participants; supervision, Y.L., Y.S. and Q.Z. All authors have read and agreed to the published version of the manuscript.

Funding: This research was funded by the National Key Research and Development Program (grant number 2016YFC0600210) and the China Geological Survey Project (grant number DD20230310).

Data Availability Statement: Data are available on request from the authors. The data are not publicly available due to confidentiality.

Acknowledgments: The authors would like to thank the anonymous reviewers and editors for their valuable suggestions.

Conflicts of Interest: The authors declare no conflicts of interest.

References

1. Zhdanov, M.S.; Wan, L.; Jorgensen, M. Joint Three-Dimensional Inversion of Gravity and Magnetic Data Collected in the Area of Victoria Mine, Nevada, Using the Gramian Constraints. *Minerals* **2024**, *14*, 292. [CrossRef]
2. Meng, Q.F.; Ma, G.Q.; Li, L.L.; Wang, T.T.; Li, Z.R.; Wang, N.; Li, J.G. Joint Inversion of Gravity and Magnetic Data with Tetrahedral Unstructured Grid and its Application to Mineral Exploration. *IEEE Trans. Geosci. Remote Sens.* **2023**, *61*, 5915514. [CrossRef]
3. Essa, K.S.; Abo-Ezz, E.R.; Anderson, N.L.; Gomma, O.A.; Elhussein, M. Magnetic inversion approach for modeling data acquired across faults: Various environmental cases studies. *Acta Geophys.* **2024**, *72*, 1819–1845. [CrossRef]
4. Essa, K.S.; Mehanee, S.A.; Elhussein, M. Gravity data interpretation by a two-sided fault-like geologic structure using the global particle swarm technique. *Phys. Earth Planet. Inter.* **2021**, *311*, 106631. [CrossRef]
5. Mehanee, S.; Essa, K.S.; Diab, Z.E. Magnetic data interpretation using a new R-parameter imaging method with application to mineral exploration. *Nat. Resour. Res.* **2021**, *30*, 77–95. [CrossRef]
6. Essa, K.S.; Mehanee, S.; Elhussein, M. Magnetic Data Profiles Interpretation for Mineralized Buried Structures Identification Applying the Variance Analysis Method. *Pure Appl. Geophys.* **2021**, *178*, 973–993. [CrossRef]
7. Prutkin, I.; Vajda, P.; Jahr, T.; Bleibinhaus, F.; Novák, P.; Tenzer, R. Interpretation of gravity and magnetic data with geological constraints for 3D structure of the Thuringian Basin, Germany. *J. Appl. Geophys.* **2017**, *136*, 35–41. [CrossRef]
8. Wei, D.G.; Jie, Y.J. Characteristics of Regional Geological Structures in Fujian Province. *Reg. Geol. China* **1997**, *16*, 162–171.
9. Yuan, H.X.; Chen, H.H. Progress in mineral exploration and characteristics of gold mineralization in Dehua area, Fujian. *J. Mineral.* **2015**, *35*, 1055.
10. Li, G.K. The relationship between the basic framework of Fujian's tectonic structure and the distribution of metallic minerals. *Reg. Geol. China* **1982**, *1*, 55–65.
11. Xiang, H.L. The controlling factors and genesis of the Qiuchun gold deposit in Fujian. *Gold Geol.* **2002**, *1*, 40–43.
12. Zhou, W.Y. Geological characteristics, ore-forming conditions, and ore-forming laws of gold deposits in the Youxi Dehua Yongtai area of central Fujian. *Geol. Fujian* **1996**, *15*, 199–208.
13. Bao, S.Z. Discussion on the ore control conditions and prospecting direction of Fujian Yan Gold Mine. *Fujian Geol.* **2015**, *34*, 289–297.
14. Mao, J.W.; Xie, G.Q.; Li, X.F.; Zhang, C.Q.; Mei, Y.W. Mesozoic large scale mineralization and multiple lithospheric extension in south China. *Earth Sci. Front.* **2004**, *11*, 46–56.
15. Hua, R.M.; Chen, P.R.; Zhang, W.L.; Lu, J.J. Three major metallogenic events in Mesozoic in South China. *Miner. Depos.* **2005**, *24*, 99–107.
16. Yue, L.Q.; Huang, C.P.; Zhang, K.Y. Discussion on the metallogenic epoch of gold deposits in Fujian province, southeast China. *Geol. Prospect.* **2002**, *38*, 17–21.
17. Xu, N. A Study on the Metallogenic System and Genesis of Dongyang Gold Mine in Fujian Province. Ph.D. Thesis, China University of Geosciences, Beijing, China, 2017.
18. Xu, Q.H.; Xie, W.Q.; Chen, Y.R. A Comprehensive Geochemical Anomaly Zoning Model and Exploration Prediction Indicators for Qiuchun Gold Mine in Fujian Province. *Geol. Explor.* **2005**, *1*, 56–61.
19. Lei, F.Z. A Study on the Genesis of Shallow Hydrothermal Gold Deposits in Dongyang, Fujian Province. Ph.D. Thesis, Jilin University, Changchun, China, 2019.
20. Huang, C.P. Research on the Metallogenic Characteristics and Prediction of Gold Deposits in Fujian Province. Ph.D. Thesis, China University of Geosciences, Beijing, China, 2017.
21. Zhang, Z.; Hu, X.; Fang, H.; Wang, D. Study on Deep Geological Structure in Dongyang Area of Fujian Province, and Provide Theoretical Basis for Deep Prospecting Breakthrough. In *AGU Fall Meeting;* The SAO Astrophysics Data System: San Francisco, CA, USA, 2019.
22. Li, G.K.; Liang, S.J. Regional geological characteristics of Fujian province. *Geol. Bull. China* **1987**, *3*, 201–210.
23. Wang, J.F. Preliminary opinions on the geological structural characteristics of Fujian and the ore-forming relationship of magmatic activity. *Jiangxi Build. Mater.* **2016**, *12*, 236.

24. Zhang, X.X. A Study on the Formation and Structural Deformation Evolution of the Neoproterozoic Mamianshan Group in the Central Fujian Region. Ph.D. Thesis, China University of Geosciences, Beijing, China, 2017.
25. Chen, D.W. Geological characteristics of the Shuangqishan gold deposit in Fujian and analysis of its potential for prospecting in the surrounding areas of the mining area. *Geol. Fujian* **2012**, *31*, 1–11.
26. Zhu, Y.L.; Liu, Y.F.; Zhang, J.T. Geological characteristics and genesis of Xiaban porphyry gold deposit in Dehua, Fujian. *Geol. Fujian* **2009**, *28*, 21–35.
27. Lei, F.Z.; Sun, J.G.; Xu, Z.T.; Han, J.L.; Liu, Y. A Study on the Metallogenic Process and Genesis of the Dongyang Shallow Hydrothermal Gold Deposit in Dehua, Fujian Province. *World Geol.* **2019**, *38*, 339–353.
28. Lin, Z.L. Geological characteristics and genesis of gold deposits in the Fufenglin mining area of Dehua, Fujian. *Geol. Fujian* **2019**, *38*, 208–215.
29. Duan, G.; Chen, H.; Xu, C.; Qi, J. Metallogenic Model of Porphyry Shallow Temperature Superimposition in the Shandong South Oreblock of Zijin, Fujian Province. *J. Mineral.* **2015**, *35*, 12.
30. Mao, G.W.; Tan, Y.S.; He, D.H.; Nie, D.P.; Liu, X.D. Exploration of mineralization types and prospecting directions in the Jinban ore section of the Taihua Mountain gold deposit in central Fujian. *Depos. Geol.* **2017**, *36*, 736–748.
31. Wang, X.Z. Geological characteristics and ore-forming prediction of the Xiaoban East Belt gold deposit in Youxi, Fujian. *Energy Environ.* **2016**, *2*, 112–113.
32. Ren, Z.Y.; Liu, J.X.; Yue, G.X.; Zhang, Y.L. Research progress of gravity and magnetic methods in the exploration of non-ferrous metal minerals. *Chin. J. Nonferrous Met.* **2023**, *33*, 240–260.
33. Liu, H.L.; Chen, C.D.; Fan, J.J.; Zhang, X.; Peng, W.X.; Zhang, W. Fault structure identification based on gravity data and its application in the Wuwei area. *World Geol.* **2023**, *42*, 334–342.
34. Wijns, C.; Perez, C.; Kowalczyk, P. Theta map: Edge detection in magnetic data. *Geophysics* **2005**, *70*, L39–L43. [CrossRef]
35. Shen, H.Y. Research on Computer Automatic Identification and Zoning Technology for Regional Magnetic Anomaly. Master's Thesis, China University of Geosciences, Beijing, China, 2019.
36. Yang, J.; Wang, Q.; Guo, J.; Zeng, Q.Q. Wavelet modulus maximum decomposition of gravity and magnetic anomalies and its application in Beiya mining area. *Geophys. Geochem. Explor. Calc. Technol.* **2019**, *41*, 485–494.
37. Wei, M.J. Determination of Normal Field Values in Magnetic Exploration. *China Sci. Technol. Inf.* **2012**, *23*, 35–40.
38. Guo, L.H.; Meng, X.H.; Shi, L.; Chen, Z.X. Optimized Filtering Method and Its Application to Bouguer Gravity Anomaly Data Processing in Chinese Mainland. *J. Geophys.* **2012**, *55*, 4078–4088.
39. Yao, C.L.; Li, H.W.; Zheng, Y.M.; Meng, X.H.; Zhang, Y.W. Comprehensive Analysis and Research on Iterative Method in Calculation of Gravity and Magnetic Potential Field Conversion. *J. Geophys.* **2012**, *55*, 2062–2078.
40. Zhang, F.X. Research on High-Precision Gravity Anomaly Data Processing Methods and Technologies. Ph.D. Thesis, Jilin University, Changchun, China, 2006.
41. Li, L.; Wu, Y.G.; Yang, C.B.; Huan, H.F.; Guo, C.C. Several boundary identification methods of gravity data and application in Vientiane of Laos area. *Glob. Geol.* **2014**, *17*, 55–61.
42. Gordon, G.R.J.; Duncan, D.R. Edge enhancement of potential-field data using normalized statistics. *Geophysics* **2008**, *71*, H1–H4.
43. Xu, S.Z. Integral Iterative Method for Potential Field Extension. *J. Geophys.* **2006**, *4*, 1176–1182.
44. Zhang, Z.H.; Wu, L.Y. The correlation coefficient method for downward extension of potential field. *J. Jilin Univ. (Earth Sci. Ed.)* **2012**, *42*, 1912–1919.
45. Gao, Y.W.; Luo, Y.; Wen, W. Research and application of compensation downward extension method. *J. Geophys.* **2012**, *55*, 2747–2756.
46. Jia, Z.; Li, Y.S.; Wang, Y.H.; Li, Y.; Jin, S.B.; Li, Y.X.; Lu, W.K. Deep Learning for 3-D Magnetic Inversion. *IEEE Trans. Geosci. Remote Sens.* **2023**, *61*, 5905410. [CrossRef]
47. Utsugi, M. 3-D inversion of magnetic data based on the L1–L2 norm regularization. *Earth Planets Space* **2019**, *71*, 73. [CrossRef]
48. Li, Y.; Oldenburg, D.W. 3-D inversion of magnetic data. *Geophysics* **1996**, *61*, 394–408. [CrossRef]
49. Wang, W.Y.; Qiu, Z.Y.; Yang, Y.; Shi, W.J. Research progress in edge recognition methods for potential fields. *Prog. Geophys.* **2010**, *25*, 196–210.
50. Ma, L.; Meng, J.H.; Shan, Z.X.; Li, X.L.; Li, F.Y.; Jin, Y.M.; Wang, J.H.; Li, B.L. Comparison and application research on new methods for identifying the boundaries of gravity and magnetic anomaly sources. *Prog. Geophys.* **2017**, *32*, 2514–2519.

minerals

MDPI

Article

Airborne Natural Total Field Broadband Electromagnetics—Configurations, Capabilities, and Advantages

Alexander Prikhodko *, Andrei Bagrianski and Petr Kuzmin

Expert Geophysics Limited, Aurora, ON L4G 1G2, Canada
* Correspondence: alexander@expertgeophysics.com

Abstract: The airborne electromagnetic system MobileMT exploits natural fields in a broadband frequency range with offset measurements of magnetic and electric field variations. It was introduced in 2018 and has since been developed in various configurations, each tailored to meet the demands of different exploration tasks, varied terrains, and geoelectrical conditions and support time-domain data with controlled primary field sources. There are four distinct airborne systems: the original MobileMT; the lighter configuration, MobileMTm; the configuration for a drone carrier, MobileMTd; and the innovative time-domain AFMAG hybrid, TargetEM. The paper describes the technical features of each system, their differences and inherent strengths, the optimal usage conditions, and insights into their applications under different conditions across various exploration tasks. Several field case studies are provided to support the natural field electromagnetics capabilities of recovering geological structures in a wide depth range, beginning from the near surface, and address the impact of parasitic IP effects on time-domain data.

Keywords: airborne EM; natural field; mineral exploration; MobileMT

1. Introduction

Airborne electromagnetic (EM) systems used in mineral exploration are based on various principles and technical designs, leading to variable performance under different geoelectrical and geological conditions. Systems with controlled primary fields have limitations, particularly in the depth of investigations and the resistivity detectability range (the last in the time-domain method). These limitations are overcome by using "passive" or natural field systems, as confirmed by direct comparisons of "passive" field data (MobileMT) with airborne TDEM data [1]. A brief history of natural field airborne technology development is described by the authors of [2], with a comparison of different systems' technical specifications and their evolution over sixty years. These systems include the original AFMAG used in the 1960s and early 1970s, further experimental prototypes of the AFMAG method (late 1990s–early 2000s), and the magnetovariational tipper-type ZTEM system patented in 2005. The latest development, MobileMT, was introduced to the airborne geophysics market in 2018 [3,4]. It measures natural magnetic (magneto) and electric (telluric) field variations across a wide range of frequencies divided into comparatively narrow windows. Using three orthogonal components to measure magnetic field variations, the system can detect geoelectrical boundaries in any direction, improving subsurface recovery in complex geological structures. The airborne technology is versatile and can be used for various geophysical tasks in diverse geoelectrical and terrain conditions, as well as different survey configurations. Technical adaptations are required for specific conditions, such as using lightweight systems at high altitudes, ensuring precise positioning for detailed surveys, addressing time-domain data insufficiencies, and exploring deeper beneath conductive overburden.

2. Theory and the System Configurations

The operating principle of airborne natural field MobileMT EM technology is a combination of magnetotelluric (MT) and magnetovariational (MV) concepts [2]. The measuring system for all configurations includes two main parts (Figure 1):

- Three orthogonal dB/dT inductive coils (Figure 1b) in a teardrop-shaped shell towed below the helicopter. Variations in the measured magnetic field (H-field) are recorded digitally in an acquisition system placed inside a helicopter. It is unnecessary to monitor or control the tilt precisely because the measurement system provides rotationally invariant total-field data;
- Two pairs of independent grounded orthogonal (X and Y) electric lines (Figure 1a) measure "signal" and "reference" variations in the electric field (E-field). Uncorrelated variations in the E-field, measured separately by the "signal" and "reference" lines, are used to clean the electrical component data [5], assuming the noise is uncorrelated with the signal. This process, using the cross-spectral technique, significantly reduces the risk of biased results [6]. The data from the stationary E-field measurement system are recorded by a separate acquisition system at the same sampling rate as the mobile H-field.

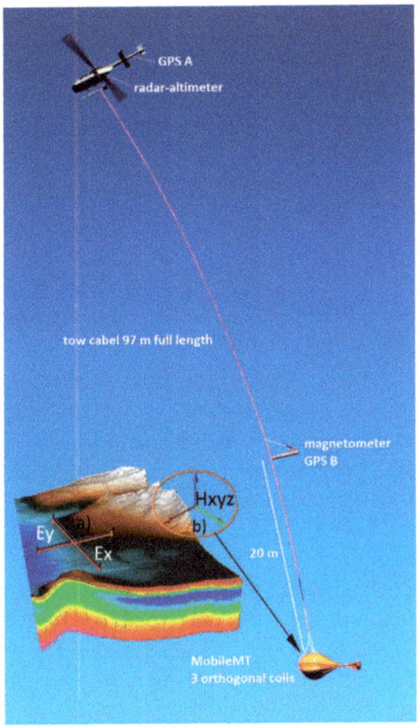

Figure 1. MobileMT system in survey configuration. (**a**) Schematic of a base station that includes two pairs of independent grounded orthogonal electric lines in the same position. (**b**) Schematic of three orthogonal dB/dT inductive coils.

The denoised and corrected E-field data represent the primary natural electromagnetic field variations. They facilitate the separation of the time variance from the space variance of the measured fields (like in MV). The combination of magnetic (H) and electric (E) fields variations are used for the admittance tensor calculation, described by Bostick and Smith [7] as Y = H/E (written in tensor notation) and, ultimately, for the calculation of apparent conductivities corresponding to different frequency bands:

$$\sigma(\omega) = \mu\omega \mid Y^2 \mid,$$

where μ is the magnetic permeability of free space, and ω is the angular frequency.

With magnetic and electric field data variations measured in different relative orientations, the magnitudes of the total H and E vectors independent of the sensors' spatial attitudes are calculated at the same frequency and time [5]. The processed data typically span a frequency range of 26–21,000 Hz, divided into 30 windows (Figure 2). The dead-band range, shown as typical in the figure, varies diurnally and seasonally.

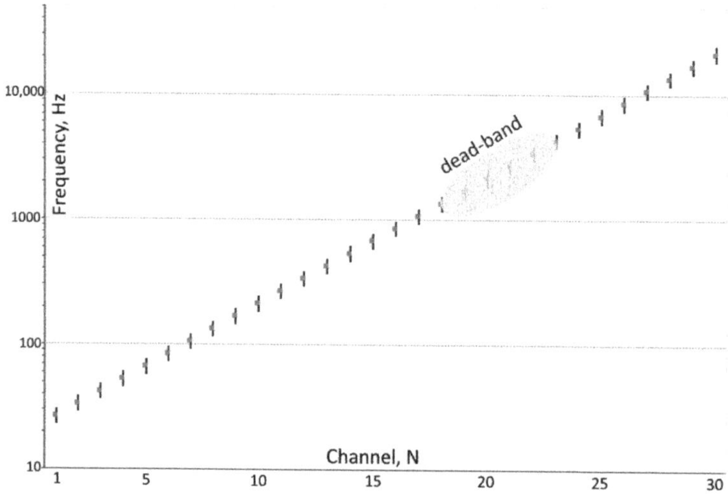

Figure 2. Typical MobileMT data frequency windows with the frequency zone (dead band), where the natural signal strength attenuates to a small level.

Currently, there are four modifications to the MobileMT system, tailored to specific survey requirements or terrain conditions:

(1) The basic model (Figure 1: 1.4 m diameter coils and 97 m tow cable) can provide data in the 26–21,000 Hz bandwidth. The historical lowest acceptable frequency data from the configuration is 22 Hz [5]. The system exhibits minimal mechanical noise, with negligible disturbance from helicopters, and is free from nearby noise sources. A GPS antenna with a Cs magnetic sensor is located 20 m above the magnetic field variations receiver in a separate bird. The system weighs 250 kg;

(2) MobileMTm (Figure 3: 0.7 m diameter receiver coils and 55 m tow cable length). Currently, the recorded frequency range is 50–28,000 Hz. Two Cs magnetic sensors in the horizontal gradiometer configuration (4 m apart), a gyro inclinometer for the magnetic sensors tilt corrections, and a GPS antenna are located on the same frame together with the three components of the magnetic variations receiver. The system weighs 150 kg.

A GPS antenna and a gyro inclinometer on the MobileMTm frame allow accurate retrieval of positions for both the airborne EM sensor and magnetometers. The precise positioning suits detailed surveys with relatively small-line spacing focusing on near-surface targets along the recovering deep structures. In addition, the system's light version (MobileMTm) perfectly matches the surveys' requirements at high altitudes >4000 m above sea level;

(3) The AFMAG component can be derived from streaming data recorded during surveys using the time-domain (TEM) system TargetEM when the MobileMT base station is operational and captures variations in the electric field (Figure 4) [1]. The receiver coils have a diameter of 1 m and are attached to a 50 m long tow cable. In the case of TEM combinations, the extracted natural field frequency range is determined by the base frequency of the controlled primary field source and the current waveform duty

cycle. Typically, the TargetEM system operates at a base frequency of 25/30 Hz, and apparent conductivities are calculated from streaming EM data in the high-frequency range of 5000–28,000 Hz, depending on the natural signal;

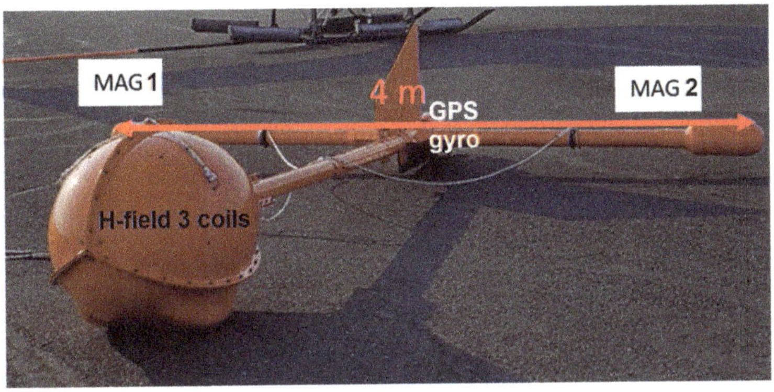

Figure 3. Airborne component of MobileMTm system.

(a) (b)

Figure 4. TargetEM system (**a**), with the ground E-field base station acquisition system (**b**).

(4) MobileMTd is a drone version of the MobileMT system (Figure 5) currently undergoing field tests designed to measure magnetic field variations at lower frequencies of 10-15-20 Hz by mitigating motion noise within this range. These lower frequencies are essential for exploring conductive areas and regions with thick, conductive overburden where the standard 26 Hz system reaches its limits. As illustrated in Figure 6, the system aims to enhance the depth of investigation (DOI), particularly in conducive environments. The DOI typically refers to the sensitivity of acquired data to subsurface petrophysical variations. Spies [8] suggests that the magnetotelluric method can detect a buried halfspace beneath 1.5 skin-depths of overburden. For instance, a 1.5 skin-depth corresponds to 600 m at 5 ohm-m in an infinite halfspace (Figure 6). Moreover, the MobileMTd system offers flexibility in selecting optimal survey times, including during and after sunset, to maximize natural electromagnetic activity peaks.

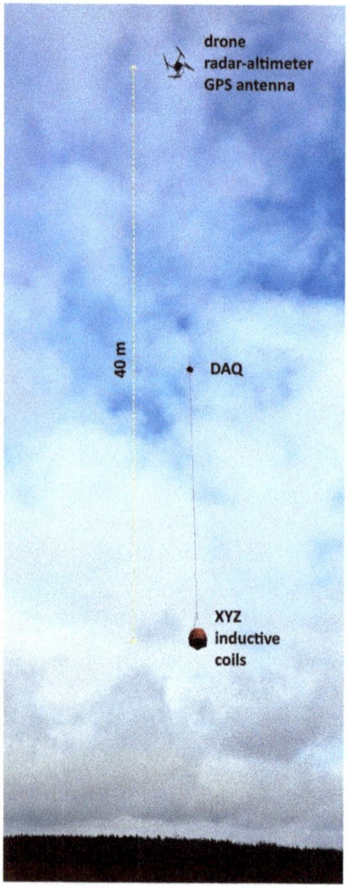

Figure 5. MobileMT system on a drone.

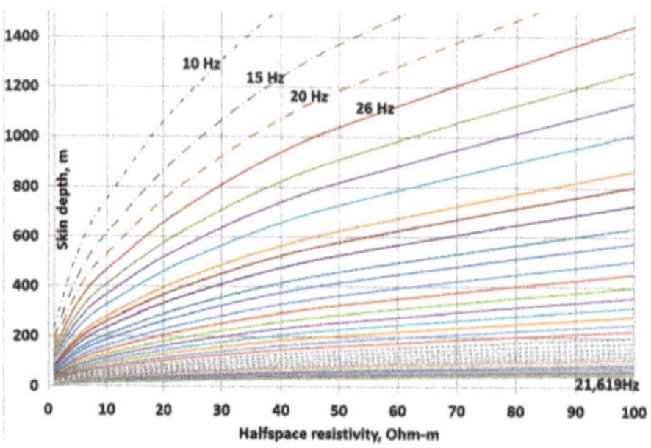

Figure 6. Diagram of 1.5 skin-depth with MobileMTd additional frequencies (10–20 Hz) calculated for conductive halfspace of 1–100 ohm-m.

In addition to natural EM field data, all configurations of MobileMT airborne technology can measure VLF data, extracting the signal from the same magnetovariational streaming data. The VLF total field magnitudes (usually in the 15–30 kHz range) are calculated as a vector of signals from the orthogonal receiver coils.

The natural total field airborne EM system, with its comprehensive technical solutions, effectively explores various types of mineralization across a wide depth range and identifies complex geological settings and structures. Table 1 describes the main technical features of MobileMT technology and their corresponding outcomes.

Table 1. Technical solutions and their outcomes in the system's capabilities.

Technical Solution	Outcome
Primary field: naturally occurring subsurface electromagnetic plane wave.	Depth of investigation consistently exceeds the capabilities of controlled-source airborne EM systems. There is no critical dependence on the system's terrain clearance, as illustrated in [5].
Three orthogonal receiver coils (total field).	Sensitivity to geoelectric boundaries in any direction.
Remote signal-reference electrical component station in the combination of a mobile magnetic component.	Denoised and bias-free data related to the electromagnetic admittance with the calculation of the absolute values of conductivities.
Frequency domain data.	Sensitivity in a full range of rock and mineral resistivity. The method is sensitive to conductors and resistivity differences in the range of thousands and tens of thousands of ohm-m (a proven case up to 20,000 ohm-m [5]).
Broadband frequency range over 3+ decades (typically 26–21,000 Hz).	Imaging of near-surface structures as well as those at >1 km depth, depending on the conductance of the geologic environment.
Output apparent conductivity data for up to 30 different frequencies (typically for 15–24 frequencies, depending on the natural signal).	Better in-depth resolution than ZTEM, with 4–6 frequencies [2] and a good opportunity for data selection, depending on cultural noise sources, natural EM field signal, and exploration goals.

3. MobileMT Capabilities and Advantages on Field Examples

The MobileMT data inversions presented below were executed using the MARE2DEM software code [9] adopted for MobileMT data. The data inversions were executed without constraints using a uniform halfspace as an initial model.

3.1. *Athabasca Basin (Canada): Comparison with Natural Field Airborne System ZTEM (Tipper Data) and Ground TAMT*

Athabasca Basin has historically presented a challenging environment for testing geophysical technologies, including airborne electromagnetic (EM) systems. This unique geological region offers a practical testing ground for several critical parameters: depth of investigation, resistivity resolution (including in high-resistivity bands), sensitivity to different boundary orientations, and the ability to recover complex structures and a low-contrasting contact.

The study area is located in the western part of the Athabasca Basin (Figure 7). The known unconformity-type uranium deposits (Collete, 58B, Kianna, Anne) are controlled by the NNW trending, graphite-rich "Saskatoon Lake Conductor (SLC)" [10]. The Athabasca Group sandstone with thickness between 710 m and 750 m unconformably overlies basement crystalline rocks and granitic and pelitic gneisses [11]. There are three types of uranium mineralization displayed on the Kianna deposit: unconformity mineralization associated with the conductive graphitic fault (SLC) and followed by intense chlorite–pyrite alteration; basement mineralization in steep-to-moderate dipping veins with intense clay–chlorite alteration; and above the unconformity alteration plume containing perched mineralization, followed by clay–chlorite alteration. The last "often occurs along the up-dip projection of basement-hosted faults into the sandstone column" [10].

The MobileMT system was tested in 2018 over the Shea Creek uranium deposit along a line and compared with the ZTEM system testing results [12] (Figure 7). The ZTEM system, a predecessor of MobileMT technology, measures only one vertical component of the magnetic field (Hz) along survey lines [2]. The orthogonal horizontal components of the magnetic field Hx and Hy are measured at a remote, stationary base station to reference the primary field variations. In contrast to MobileMT, ZTEM does not utilize the data acquired at the station as a reference to denoise the data, which could cause biased tipper data [13]. Other limitations of the tipper-type system include a lack of ability to image layered geology [13], a reduced bandwidth, a limited number of frequencies, and comparatively wide frequency windows [2].

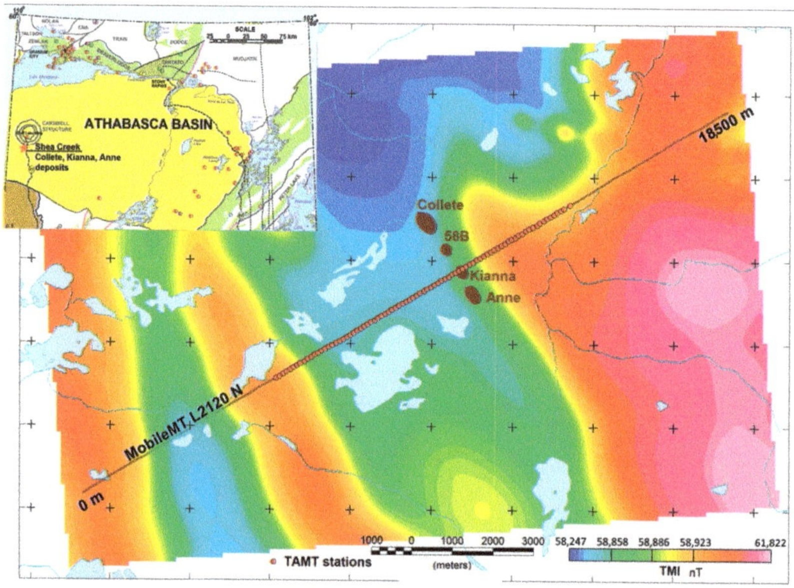

Figure 7. MobileMT (and ZTEM [12]) survey line with positions of TAMT stations on a magnetic field map of the study area. The overview geological map from the Mineral Resource Map of Saskatchewan, 2008 edition (Saskatchewan Ministry of Energy and Resources). Magnetic field data and the overlapped hydrology from Canada Geoscience Data (http://gdrdap.agg.nrcan.gc.ca, accessed on 6 May 2024).

Figure 8 compares resistivity sections derived from MobileMT and ZTEM data along the same survey line crossing the SLC with the Kianna deposit [10] and the Klarke Lake structural conductor. Apparent conductivities across twelve frequency windows used in MobileMT data inversions are displayed in the resistivity section. While both systems detect major conductive structures, they differ in performance in several aspects. Unlike ZTEM, the MobileMT system clearly identifies the unconformity contact between the more resistive Athabasca sandstones and basement rocks at a depth of 700–750 m below the surface. On the right side of the survey line, ZTEM depicts a continuous conductive layer at the top, ranging from 700 to over 1000 m thick, interpreted as Douglas Formation mudstones [12]. According to the descriptions of stratigraphy and sedimentology of the western Athabasca Basin, the thickness of Douglas Formation mudstones typically does not exceed 200 m [14]. MobileMT provides a more detailed depiction in this segment, revealing a near-surface conductive layer with a thickness of approximately 200 m (likely corresponding to the Douglas Formation mudstones) and a distinct conductor in the basement.

Data from the ground transient audio magnetotellurics (TAMT) method exploiting linearly polarized signal of sferics [15] were collected in the summer of 2005 over the Shea Creek deposit [16]. One of the ground survey lines crosses the Kianna mineralization zone and lies in the central part of the MobileMT and ZTEM test line, as shown in Figure 7. The TAMT resistivity section extracted from a 3D model of the tipper data is shown in Figure 9, along with MobileMT resistivity distribution in the same line range. The shape of the Saskatoon Lake conductor, as recovered from MobileMT data, is well-aligned with the results of the ground TAMT survey (Figure 9).

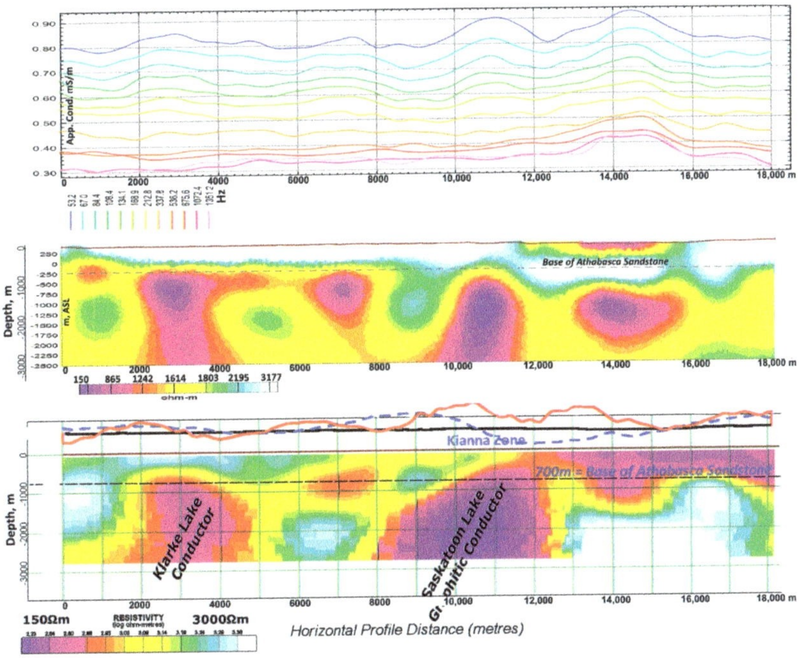

Figure 8. MobileMT apparent conductivity profiles (**top**); MobileMT resistivity section (**middle**) and ZTEM resistivity section (from [12], **bottom**) over the line crossing Kianna uranium mineralization zone.

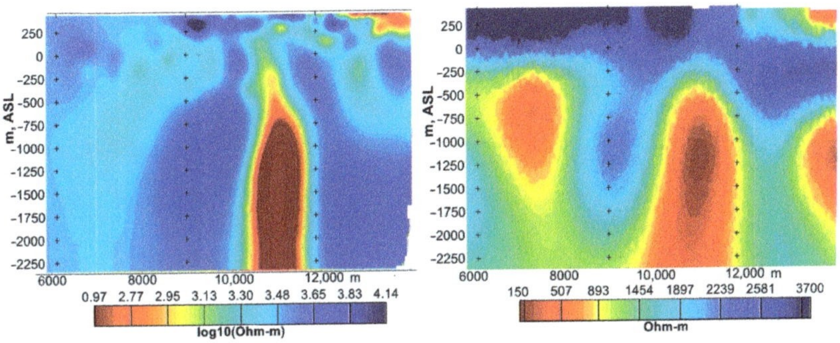

Figure 9. Kianna zone: **left**—resistivity section extracted from ground TAMT tipper (Zxy/Zyx/Tx/Ty) 3D model; **right**—MobileMT resistivity section.

3.2. VMS Mineralization System El Domo (Ecuador)

The El Domo deposit is a gold-rich, polymetallic VMS deposit located in the Western Cordillera of Ecuador. The mineralization is flat-lying, stratiform, and stratabound and occurs in one main massive sulfide lens, a directly overlying talus or breccia zone, and a number of smaller, mineralized lenses, primarily in the footwall of the main lens [17,18]. The lens thickness ranges between 20 cm and 25 m and strictly follows the contact between the Lower Felsic Unit and the Hangingwall Unit (Figure 10). Sphalerite, chalcopyrite, and pyrite are the principal sulfides in the mineralized rocks. Galena is less common, and tennantite/tetrahedrite and covellite are minor phases. The known lateral dimensions of the VMS massive sulfide mineralization are approximately 1,000x800 m. The massive sulfides are related to a zone of abundant hydrothermal alteration, which includes extensive sericitization–silicification in the rhyodacitic footwall and widespread silicification–chloritization–argillitization in the overlying mafic volcaniclastic rocks. The rhyodacite hosts a sulphide-rich stockwork zone and abundant gypsum, replacing earlier anhydrite. The stockwork is characterized by quartz–sericite alteration and includes massive pyrite mineralization, which is irregularly replaced by abundant chalcopyrite [17]. The El Domo model type showcases distinct zoning, starting from the underlying feeder pipe area (considered the stockwork) and extending through vertical and lateral variations up to the abrupt termination of the massive sulfides [17].

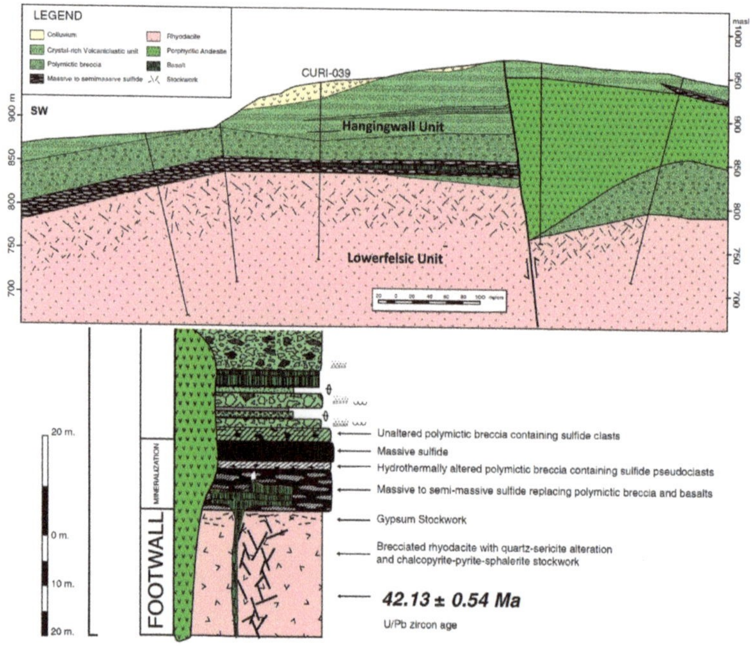

Figure 10. Geological section (**top**) and stratigraphic column (**bottom**) of El Domo deposit [18].

There is no proven evidence from historical geophysical data that the main lens of El Domo VMS mineralization is conductive compared to the surrounding rocks. If sphalerite, a non-conductive mineral, is the primary mineral in the massive sulfides assemblage, while pyrite and chalcopyrite are found disseminated or aggregated with quartz and barite, as indicated in [17], then the lens' potential for high conductivity is questionable. The MobileMT survey results suggest that the flat-lying mineralization is located on the contact between the resistive rocks of the Hangingwall Unit and the conductive footwall (Figure 11), as the drilling results indicate directly [17,18]. The conductive zone exhibits complex structure boundaries, likely indicating alteration zones associated with the stockwork controlling VMS mineralization as a feeder structure rather than uniformly representing Lowerfelsic Unit rocks. Figure 12 shows the resistivity distribution around the El Dome deposit in a 3D view. The El Domo case study demonstrates MobileMT's capabilities in recovering comparatively near-surface geology and deep structures, including complex geometry.

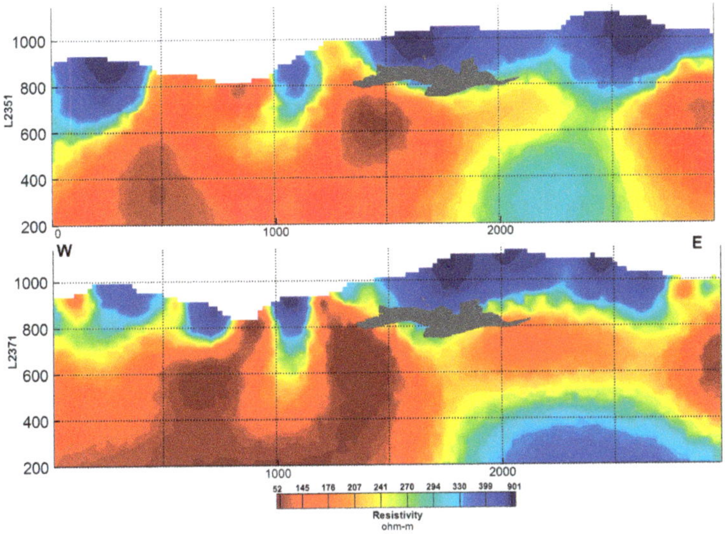

Figure 11. MobileMT resistivity sections along survey lines crossing El Domo deposit (distance between L2371 and L2351 lines is 200 m). Grey—projection of stratabound VMS mineralization on the survey lines.

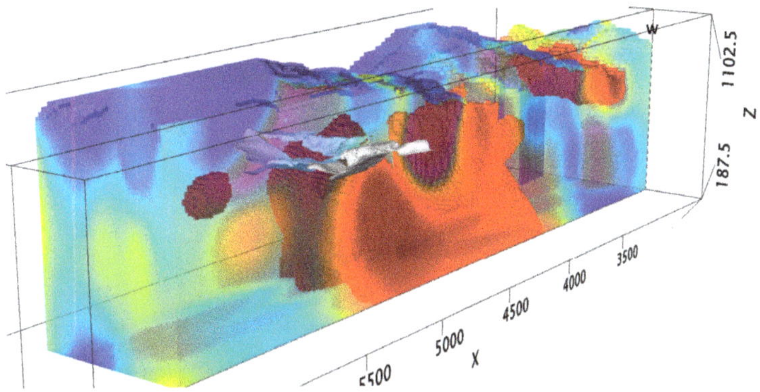

Figure 12. MobileMT resistivity 3D voxel with the El Domo VMS mineralization position (grey).

3.3. *Sudbury Impact Structure (Canada)*

One of the main environments of the Ni-Cu sulfide and platinum group element (PGE) mineralizations in the 1.85 Ga Sudbury impact structure is the near basal contact and the underlying anatectic footwall breccia [19]. The contact of the Sudbury Igneous Complex (SIC) contains a significant amount of pyrrhotite with pentlandite and chalcopyrite, but the PGE mineralization at Sudbury is not always associated within the highest concentrations of sulfide and often occurs hundreds of meters away from the Ni–Cu mineralization and within the footwall rocks [20]. The SIC footwall strata, associated breccia, and their inflections and hanging forms are important structural and lithological factors in controlling both types of mineralization in the geological structure.

Moderately conductive bodies within the norite, contact sublayer, and footwall breccia area were mapped using heliborne frequency-domain data employing coplanar (32,000 and 4175 Hz) and coaxial (4600 and 935 Hz) coil pairs [21]. Both MobileMT airborne EM technology and the frequency-domain method can detect not only highly conductive sulfide concentrations but also moderately and low-conductive structures and lithologies [2].

MobileMT operates across a broader spectrum of frequencies, enhancing its depth of investigation (DOI) compared to frequency-domain systems with controlled primary field sources. MobileMT's DOI is estimated to reach nearly 2 km in the Sudbury resistive environment. The lowest frequency used in MobileMT data inversions was 84 Hz due to interference from nearby powerlines and industrial sources affecting lower frequencies. The results from MobileMT EM data inversion along a test line crossing the SIC contact in its southwest end are presented in Figure 13, demonstrating a depth of investigation estimated to exceed 2000 m, based on a sensitivity measure of approximately −2.7 on a log10 scale (Figure 14).

The position of the line is shown on the geological map (Figure 15). The conductive footwall between the SIC and Precambrian igneous rocks has been mapped successfully. The test survey results demonstrate MobileMT's capability of detecting and recovering deep structures, which are not reachable by conventional airborne EM systems with towed controlled sources of the primary field.

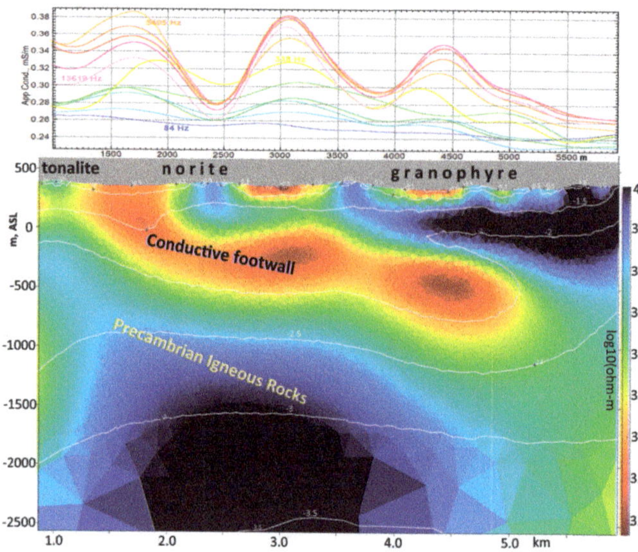

Figure 13. MobileMT apparent conductivity profiles in 84–13,619 Hz bandwidth (**top**) and resistivity section with overlapped normalized inversion sensitivity contours (the test survey line position is in Figure 15).

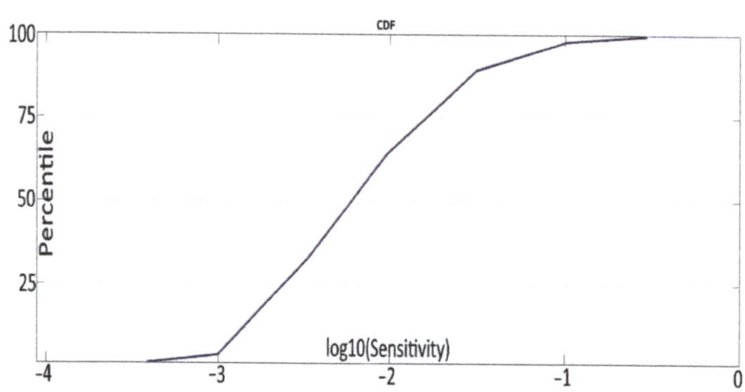

Figure 14. Cumulative distribution function of normalized inversion sensitivity along the survey line in Figure 9.

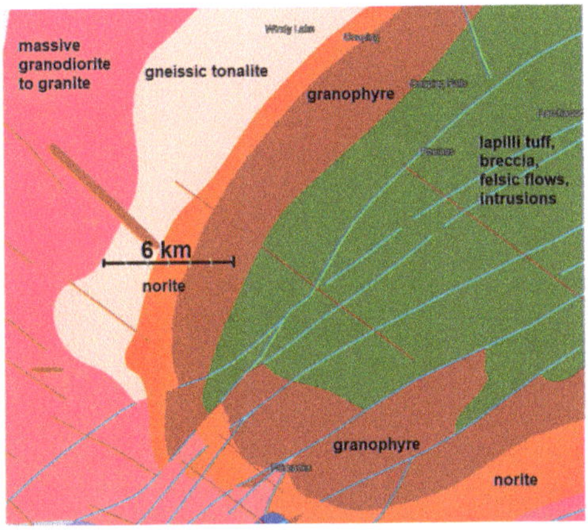

Figure 15. SW of Sudbury impact structure geology (Ontario Geological Survey).

3.4. *Poplar Porphyry Deposit (BC, Canada)*

Copper–molybdenum porphyry mineralization on the north shore of Tagetochlain Lake is associated with the Late Cretaceous Poplar intrusive stock. The sulfide mineralization occurs within broad envelopes of propylitic, argillic, phyllic, and potassic alteration [22]. Featuring a well-developed pyrite halo, the deposit appears in the MobileMT data as a discrete conductive anomaly that closely matches the deposit boundaries (Figure 16). The Late Cretaceous felsic pluton (Poplar Stock) is distinctly observed as a resistive, dome-like structure in the center of the resistivity sections.

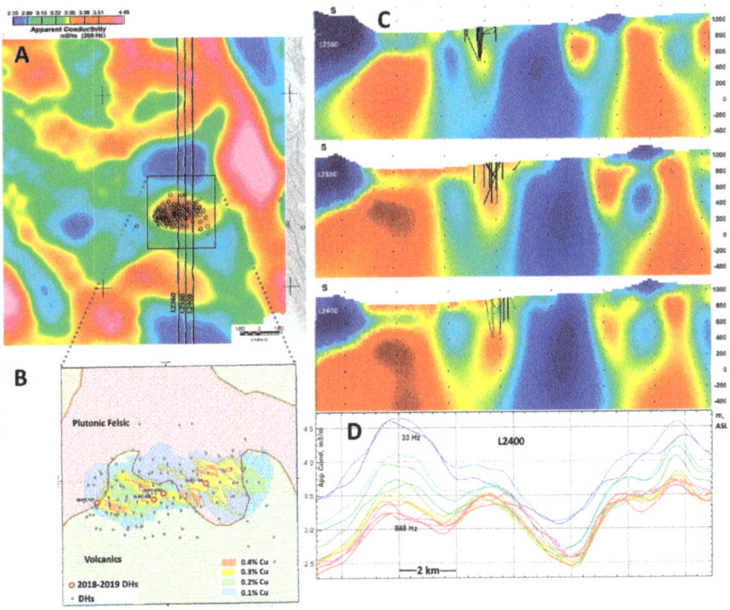

Figure 16. Poplar porphyry deposit: (**A**) Apparent conductivity color grid (266 Hz) with drill hole positions and MobileMT lines crossing the deposit; (**B**) Geological map and Cu grades projected to the surface (from [22]); (**C**) MobileMT resistivity sections along the lines in A, with projections of drill holes; (**D**) MobileMT apparent conductivity profiles along Line 2400.

3.5. Combination with Active Source Time-Domain Data

The TargetEM system was designed to capture both time-domain responses from a controlled-source transmitting field and natural (passive) EM field data [1]. Natural EM field data are recorded between the pulses of the transmitting field and when the transmitter is off as a second option. This combined active and passive airborne electromagnetic system collects broadband streaming data, allowing the extraction of variations in the natural EM field, VLF signals, and time-domain components. Even within a limited frequency range (typically above 5 kHz), natural field data play a crucial role in filling gaps where the time-domain method may be restricted, such as in mapping highly resistive geological terrains, detecting superconductors, conducting surveys in rugged terrain, and identifying parasitic effects, like IP and SPM.

An illustrative field example of TargetEM data comes from Western Australia, specifically within the Norseman-Wiluna Belt of the Kurnalpi Terrane in the Eastern Goldfields Superterrane (EGS). This region is known for significant occurrences of nickel sulfides associated with Archaean greenstone peridotites [23].

TargetEM time-domain data were recorded with a base frequency of 25 Hz and a 420,000 NIA dipole moment, covering 37 time gates within the off-time range of 83.4–12478.30 microseconds. A prominent parasitic IP effect, evident from negative dB/dt responses, significantly influenced induction, particularly in the eastern part of the survey block (Figure 17). Passive EM field and VLF data were extracted from the same streaming data recorded during the survey at a sampling rate of 73,728 Hz. As depicted in Figure 17, discrete conductors were identified within the IP anomaly using complementary high-frequency data extracted between transmitting field pulses.

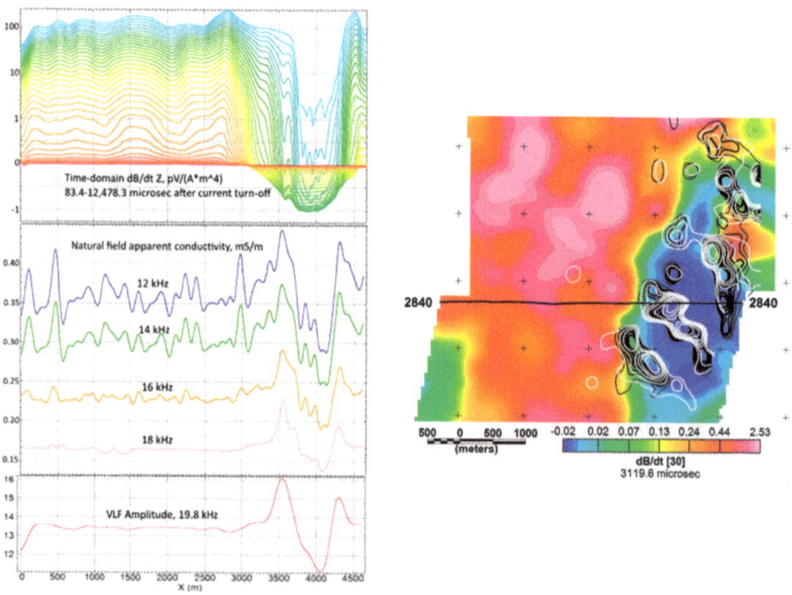

Figure 17. Left: TargetEM time-domain, natural field EM, and VLF data profiles recorded simultaneously along 2840 survey line; right: dB/dt color grid with overlapped anomaly contours apparent conductivity natural field at 18 kHz (white) and VLF at 19.8 kHz (black) (Western Australia).

Furthermore, a comparison between natural field MobileMT data and time-domain data affected by induced polarization due to a superficial clay-rich layer [1] highlights the relative independence (or minimal influence) of MobileMT's natural field data on parasitic effects.

Gasperikova and Morrison [24] demonstrate that natural field (MT) data can be affected by the IP effect only under specific conditions. First of all, the IP component is prominent when telluric (E) field data are measured at a fixed base and along stations during continuous profiling (the TM mode), as E-field data become frequency-dependent in the presence of a polarizable body, indicating the presence of the IP effect. In MobileMT surveys, E-field data are measured only at a stationary position, not over potentially polarized bodies. Favorable conditions to detect the IP influence in natural field data include having a finite polarized body and low frequencies (typically below 1 Hz, outside the MobileMT frequency range), where the body shows negligible induction but still exhibits significant complex frequency-dependent resistivity. Therefore, based on Gasperikova and Morrison's research results and MobileMT survey practices, including measurements over strongly polarized surficial layers [1], MobileMT data are not affected by the IP phenomenon.

4. Discussion

The introduction of MobileMT airborne electromagnetic (EM) technology in 2018 marked a significant advancement in mineral exploration with airborne EM. It leverages natural field measurements to overcome the limitations of traditional controlled-source systems and previously developed natural field systems.

The airborne modification of the magnetotelluric method does not contradict the definition of the method given by Louis Cagniard in 1953, where the measurement of variations in magnetic and electric fields at the same positions (stations) is just a preferable option [2]. The telluric method, which involved the comparison of only horizontal electric fields measured simultaneously at a base-fixed station and remote survey sites, was used from 1939 to 1973 (before and after Cagniard–Tikhonov's discovery) by Schlumberger, Berdichevskiy, and Yungul [25]. The fundamental point of Cagniard–Tikhonov's discovery is the canceling of variations from far-zone natural sources in measured data by the mutual normalization of telluric and magnetic field components [26], and this principle is at the core of MobileMT technology. In the ground telluric–magnetotelluric method, the telluric transfer tensor (T) between the electric fields measured at different stations (r and b) is expressed as $E^r = [T]*E^b$ [25], and similarly, the magnetotelluric tensor in MobileMT is expressed as $H^r = [T]*E^b$ [5]. Combining the offset E-field and H-field in the resulting output data, as well as the total field measurements, requires modifying the standard inversion codes, as noted in [13], and several consulting groups have already achieved this. The 3D inversion of MobileMT data is implemented in EMvision (TecnoImaging), E3dMT (UBC-GIF), GeoTools (Viridien), and MAGNUM (Geotexera) software, to name a few examples.

Airborne natural field technology has several advantages over ground methods. The advantages include high-density coverage of comparatively large territories in a reasonable time, encompassing high-altitude regions, mountainous landscapes, deserts, and other hard-to-reach areas. Additionally, there are significant economic advantages. The versatility and depth of investigation offered by MobileMT and its configurations (MobileMTm, MobileMTd, and TargetEM) have enabled the detailed exploration of complex geological environments, even in challenging geoelectrical and terrain conditions. The airborne MT concept has been verified experimentally, as demonstrated in the field examples above and other publications [1–3,5]. The results from these case studies, including direct comparisons with other airborne and ground methods and drilling, highlight the broad applicability of MobileMT technology and its configurations in mineral exploration.

The following implications and future research directions are suggested:

- Enhanced subsurface imaging: MobileMT's ability to measure total natural fields over a broad frequency range allows for deeper penetration and more detailed subsurface imaging in a wide range of resistivities. Future research should focus on refining inversion algorithms and data processing techniques to further enhance the resolution and accuracy of subsurface models;

- Adaptability to diverse environments: The different configurations of MobileMT systems cater to specific survey needs, making them suitable for a wide range of exploration scenarios. Further field tests, particularly with the drone-based MobileMTd, are necessary to validate its performance and extend its applicability in challenging terrains, including very conducive terrains;
- Integration with other electromagnetic methods: Combining MobileMT with other geophysical EM techniques, such as time domain and VLF, extracted from the same streaming data, could provide a more comprehensive understanding of the geoelectrical image of the subsurface from very near-surface to the first kilometers depth. Future research and developments should explore the synergies between these methods to improve integrated exploration strategies with the EM methods in one system;
- Mitigating time-domain and natural field method data limitations: The problematic situations of the airborne time-domain method include the development of parasitic IP and SPM effects, highly resistive geology, and high system altitude in rugged relief conditions. With its comparatively large footprint and broad high-frequency windows, the natural field method has a lower resolution in detecting near-surface features and smaller conductors than the time-domain method. The further development of combined systems that integrate passive- and active-source data will be crucial in overcoming the limitations of each method and improving the overall data quality.

Author Contributions: Conceptualization, A.P. and A.B.; methodology, P.K.; data processing software, P.K.; validation, A.P., A.B. and P.K.; formal analysis, A.P.; investigation, A.P.; resources, A.P.; data curation, A.P.; writing—original draft preparation, A.P.; writing—review and editing, A.P.; visualization, A.P.; supervision, A.P.; project administration, A.B. All authors have read and agreed to the published version of the manuscript.

Funding: This research received no external funding.

Data Availability Statement: Data are unavailable due to privacy restrictions.

Acknowledgments: We acknowledge Expert Geophysics Limited's field crews and technical support during the surveys under Mikhail Kuzmin's management and data quality control, the processors under Emily Data management, and Aamna Sirohey, a senior geophysicist, for her contributions and support in data inversions. We are grateful to Universal Copper for the permission to use their data and David Goldak (EMpulse Geophysics) for providing a TAMT digital 3D model for the comparative analysis. We sincerely thank the peer reviewers for their valuable insights and constructive feedback, which significantly enhanced the quality of this manuscript.

Conflicts of Interest: Alexander Prikhodko, Andrei Bagrianski and Petr Kuzmin are employed by Expert Geophysics Limited, Canada. The paper reflects the views of the scientists and not the companies.

References

1. Prikhodko, A.; Bagrianski, A.; Kuzmin, P.; Carpenter, A. Passive and active airborne electromagnetics—Separate and combined technical solutions and applicability. In Proceedings of the Extended Abstracts—8th International Airborne Electromagnetics Workshop, Fitzroy Island, Australia, 3–7 September 2023.
2. Prikhodko, A.; Bagrianski, A.; Kuzmin, P.; Sirohey, A. Natural field airborne electromagnetics—History of development and current exploration capabilities. *Minerals* **2022**, *12*, 583. [CrossRef]
3. Bagrianski, A.; Kuzmin, P.; Prikhodko, A. AFMAG evolution—Expanding limits. In Proceedings of the Extended Abstracts—16th SAGA Biennial Conference & Exhibition, Durban, South Africa, 6–9 October 2019.
4. Sattel, D.; Witherly, K.; Kaminski, V. A brief analysis of MobileMT data. In Proceedings of the 89th Society of Exploration Geophysicists International Exposition and Annual Meeting, SEG, San Antonio, TX, USA, 15–20 September 2019.
5. Prikhodko, A.; Bagrianski, A.; Wilson, R.; Belyakov, S.; Esimkhanova, N. Detecting and recovering critical mineral resource systems using broadband total-field airborne natural source audio frequency magnetotellurics measurements. *Geophysics* **2024**, *89*, WB13–WB23. [CrossRef]
6. Labson, V.F.; Becker, A.; Morrison, H.F.; Conti, U. Geophysical exploration with audiofrequency natural magnetic fields. *Geophysics* **1985**, *50*, 656–664. [CrossRef]
7. Bostick, F.S.; Smith, H.W. Investigation of large-scale inhomogeneities in the Earth by the magnetotelluric method. *Proc. IRE* **1962**, *50*, 2339–2346. [CrossRef]

8. Spies, B.R. Depth of investigation in electromagnetic sounding methods. *Geophysics* **1989**, *54*, 872–878. [CrossRef]
9. Key, K. MARE2DEM: A 2-D inversion code for controlled-source electromagnetic and magnetotelluric data. *Geophys. J. Int.* **2016**, *207*, 571–588. [CrossRef]
10. Rhys, D.; Sierd, E.; van der Meer, L. Geology of the Shea Creek uranium deposits—Part of an expanding uranium district in the Western Athabasca Basin. In Proceedings of the Saskatchewan Geological Survey Open House, Saskatoon, Saskatchewan, 29 November 2010; Volume 4.
11. Nimeck, G.; Koch, R. A progressive geophysical exploration strategy at the Shea Creek uranium deposit. *Lead. Edge* **2008**, *27*, 52–63. [CrossRef]
12. Geotech Ltd. ZTEM Tipper AFMAG Results over Shea Creek Unconformity Uranium Deposits. 2009. Available online: https://geotech.ca/wp-content/uploads/2015/02/012-ZTEM_Uranium_SheaCreek_Saskatchewan.pdf (accessed on 6 May 2024).
13. Jansen, J.C.; Cristall, J.A. Mineral exploration using natural EM fields. In Proceedings of the Exploration 17: Sixth Decennial International Conference on Mineral Exploration, Toronto, ON, Canada, 22–25 October 2017; Targeting 2: Mine to Camp Scale. pp. 349–377.
14. Post, R.T. *Stratigraphy and Sedimentology of the Athabasca Group in the Net Lake-Maybelle River Area, North-Eastern Alberta*; EUB/AGS Earth Sciences Report 2003-01; Alberta Energy and Utilities Board: Calgary, AB, Canada, 2004; 103p.
15. Goldak, D.; Goldak, M. Transient Magnetotellurics with Adaptive Polarization Stacking. In Proceedings of the SEG International Exposition and 71′st Annual Meeting, San Antonio, TX, USA, 9–14 September 2001; pp. 1509–1512.
16. EMpulse Geophysics Ltd. Case Histories, Shea Creek. Available online: http://www.empulse.ca/upload/documents/shea3d.pdf (accessed on 6 May 2024).
17. de Weerdt, P.; El Rassi, D.; Gagnon, D.M.; Bisaillon, C.; Liskovych, V.; Embree, K.; Stephens, B.; Shaw, S.; Gravel, A. NI 43-101 Technical Report Feasibility Study Curipamba El Domo Project, Central Ecuador. 2021. Available online: https://www.sedarplus.ca/csa-party/records/document.html?id=0f7ee7252be12656c6f52305d868c855f5b1023f1573e62980e1f20bc26b78da (accessed on 7 July 2024).
18. Vallejo, C.; Soria, F.; Tornos, F.; Naranjo, G.; Rosero, B.; Salazar, F.; Cochrane, R. Geology of El Domo deposit in central Ecuador: A VMS formed on top of an accreted margin. *Miner. Depos.* **2015**, *51*, 389–409. [CrossRef]
19. Wang, Y.; Lesher, C.M.; Lightfoot, P.C.; Pattison, E.F.; Golightly, J.P. Genesis of sublayer, footwall breccia, and associated Ni-Cu-platinum group element mineralization in the Sudbury igneous complex. *Econ. Geol.* **2022**, *117*, 1791–1807. [CrossRef]
20. Balch, S.J. The geophysical signatures of platinum-group element deposits. In *Exploration for Platinum-Group Elements Deposits*; Mungall, G.E., Ed.; Mineralogical Association of Canada Short Course Series; Mineralogical Association of Canada: Québec City, QC, Canada, 2005; Volume 35, pp. 275–285.
21. Olaniyan, O.; Smith, R.S.; Morris, B. Qualitative geophysical interpretation of the Sudbury Structure. *Interpretation* **2013**, *1*, T25–T43. [CrossRef]
22. Ashton, J.; Robb, W. *Technical Report on the Poplar Deposit, Omineca Mining Division*; Universal Copper Ltd.: Vancouver, BC, Canada, 2021.
23. Woodall, R.; Travis, G.A. The Kambalda nickel deposits, Western Australia. In Proceedings of the 9th Commonwealth Mining Metallurgy Congress, London Institute of Mining Metallurgy, London, UK, 29 May 29–5 June 1969; p. 26.
24. Gaspericova, E.; Morrison, H.F. Mapping of induced polarization using natural fields. *Geophysics* **2001**, *66*, 137–147. [CrossRef]
25. Hermance, J.F.; Thayer, R.E. The telluric-magnetotelluric method. *Geophysics* **1975**, *40*, 664–668. [CrossRef]
26. Zhdanov, M.S. Electromagnetic geophysics: Notes from the past and the road ahead. *Geophysics* **2010**, *75*, 75A49–75A66. [CrossRef]

minerals

MDPI

Article

Application of Integrated Geological and Geophysical Surveys on the Exploration of Chalcedony Deposits: A Case Study on Nanhong Agate in Liangshan, China

Shengping Gong [1,2], Keqiang Zhao [1,2,*], Mingming Wang [3], Shengwu Yan [4,5], Yong Li [1,2] and Jianzhou Yang [1,2,*]

1 Institute of Geophysical and Geochemical Exploration, Chinese Academy of Geological Sciences, Langfang 065000, China; gshengping@mail.cgs.gov.cn (S.G.)
2 Key Laboratory of Geophysical Electromagnetic Probing Technologies of Ministry of Natural Resources, Langfang 065000, China
3 Institute of Regional Geological Survey of Hebei Province, Langfang 065000, China
4 Sichuan Institute of Comprehensive Geological Survey, Chengdu 610081, China
5 Evaluation and Utilization of Strategic Rare Metals and Rare Earth Resources Key Laboratory of Sichuan Province, Chengdu 610081, China
* Correspondence: zkeqiang@mail.cgs.gov.cn (K.Z.); yjianzhou@mail.cgs.gov.cn (J.Y.)

Abstract: Nanhong agate, esteemed for its vivid color and natural shine, is experiencing a scarcity in supply despite its high demand. The primary deposits of agate, typically found near the surface, have not been extensively explored due to the predominance of traditional manual excavation methods. This research examined the Nanhong agate deposits in the Zhaojue–Meigu region of Liangshan, China, employing the integration of geological and geophysical surveys. Field geological surveys allowed us to outline the general areas where agate is found. Following this, using magnetic surveys, vertical electrical sounding, and controlled-source audio magnetotellurics, agate deposits were located within the conglomerate layer of the second member of the Feixianguan Formation from the Lower Triassic period at depths of less than 100 m. Our results identify mineralized layers, Xuanwei Formation mudstone, and the underlying bedrock, thus supporting the creation of a mineral prediction map. This research provides essential insights and guidance for agate exploration and the development of associated mineral resources.

Keywords: agate deposits; Liangshan; geological survey; vertical electrical sounding; controlled-source audiomagnetotelluric method

1. Introduction

The jadeite mineral group is highly esteemed in the gemstone market due to its variety of colors and textures [1,2]. Among these variants, Nanhong agate stands out for its rich crimson hue and exquisite texture, commanding significant popularity among consumers. Among these, Nanhong agate is particularly notable for its deep crimson color and fine texture, making it highly popular among buyers. Nanhong refers to a specific type of red agate found in the southwestern part of China. Notably, the Nanhong agate deposits in the Zhaojue–Meigu area of Liangshan, Sichuan Province, China, are renowned for their pristine condition [3,4]. The formation of Nanhong agate is linked with the creation of the nearby Emeishan basalt during the late Permian period. Volcanic activities during this period enabled the mineralization of agate through the percolation of silica-rich hydrothermal fluids containing metal ions, which imparted the distinctive coloration. Further geological processes, such as weathering and erosion, shaped these agate minerals, leading to the abundant presence of Nanhong agate pebbles. These pebbles were then deposited within the sedimentary layers of the Xuanwei and Feixianguan groups in the Zhaojue–Meigu area, forming secondary ore deposits [5,6].

Despite the high economic value of Nanhong agate, exploration in the region largely depends on outdated manual excavation methods. These methods are inherently inefficient and largely fortuitous, underscoring the necessity for more scientifically rigorous and efficient exploration techniques. Although geophysical methods offer considerable benefits for mineral exploration [7–12], their use in jadeite mineral investigations has been relatively limited. Therefore, this study seeks to assess the efficacy of integrated geophysical methods in the exploration of Nanhong agate deposits. We conducted a detailed and systematic exploration of Nanhong agate deposits in the Liangshan area, building on field geological surveys and employing a suite of geophysical techniques, including magnetic surveying, vertical electrical sounding (VES), and the controlled-source audiomagnetotelluric (CSAMT) method.

Geological investigations revealed that the conglomerate bed of the Lower Triassic Feixianguan Formation harbors Nanhong agate deposits, buried at depths of less than 100 m. Through the application of geophysical methods, we delineated ore-bearing layers, identified the low-resistivity mudstone of the Xuanwei Formation, and mapped the underlying bedrock. This comprehensive characterization facilitated the development of a mineral prediction map, offering insights into the precise distribution of agate-bearing layers. Our study establishes a scientific foundation for the accurate localization and evaluation of Nanhong agate deposits while also introducing innovative methodologies for jadeite mineral exploration. We anticipate that our findings will contribute to the sustainable development and utilization of Nanhong agate resources, thereby advancing exploration efforts in the domain of chalcedony mineral deposits.

2. Geological Background

The study area, located in the southwestern part of Sichuan Province, China, is part of the Emeishan large igneous province. The exposed strata primarily include the Emeishan Basalt Formation of the Upper Permian, the Xuanwei Formation, and the Feixianguan Formation of the Lower Triassic (Figures 1 and 2).

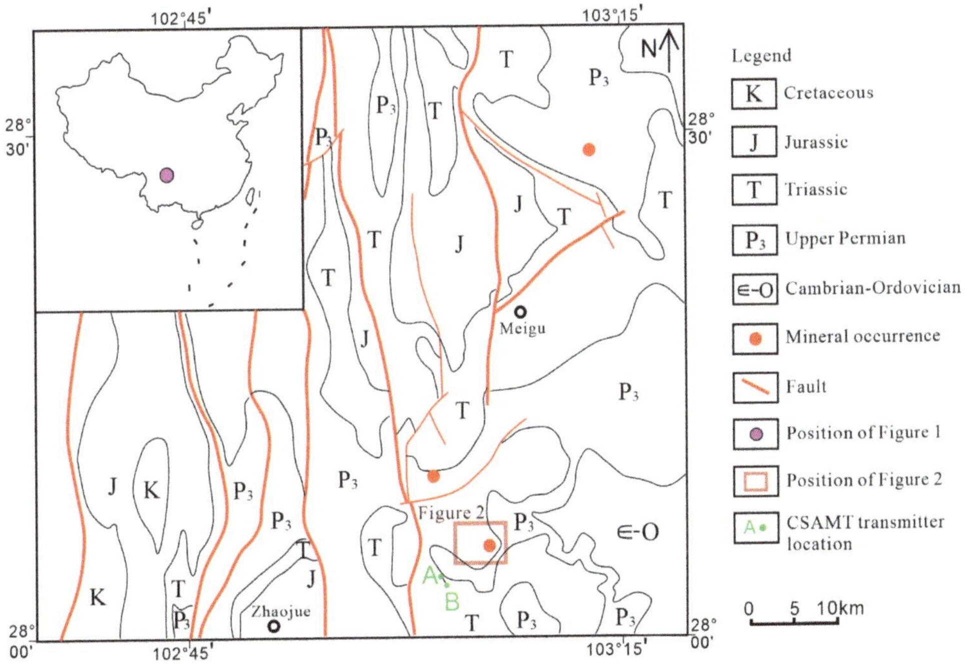

Figure 1. Simplified geological map of the study area.

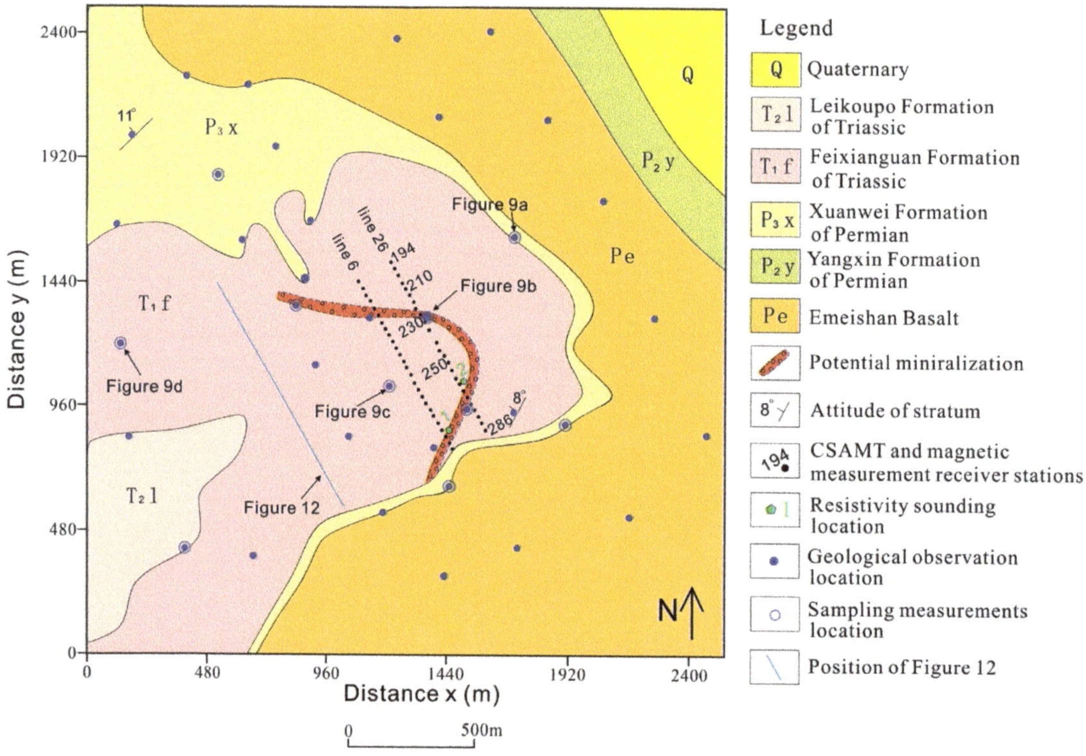

Figure 2. Geological map of the exploration area with measurement points (locations of geological and geophysical surveys are shown in the figure).

The Emeishan Basalt Formation comprises dense, mottled basalts often interbedded with volcanic breccias or tuffaceous sandstones. It nonconformably comes into contact with the underlying Yangxin Formation of the Upper Permian, with a thickness exceeding 200 m [5].

The Xuanwei Formation (P_3x) is characterized by fine-grained sandstone and iron-bearing claystone, with purple-red claystone at the base, coal-bearing mudstone in the middle, and grayish-green lithic sandstone at the top. It lies nonconformably over the Emeishan Basalt Formation, exhibiting significant weathering and sedimentary discontinuities [13].

The Feixianguan Formation (T_1f) is widespread in the study area, featuring consistent lithology with a thickness ranging from 193 to 259 m [5]. It primarily consists of purple-red gravelly feldspathic quartz sandstone and sandy shale, with gravel layers dominated by siliceous rocks and occasional agate gravel. Based on its lithological attributes, the Feixianguan Formation can be categorized into three members. The first member is characterized by thin-bedded, purple-red, medium-grained lithic sandstone alternating with purple-red thin-bedded mudstone. The second member consists of a medium–thick-bedded conglomerate displaying rhythmic layers of conglomeratic sandstone with numerous agate pebbles embedded within. Notably, sandstone layers exhibit distinct parallel bedding, tabular cross-bedding, and wedge-shaped cross-bedding. The third member comprises purple-red sandstone and fine sandstone interbedded with purple-red siltstone and mudstone. Of economic significance, agate deposits are predominantly situated within the second member of the Feixianguan Formation (Figure 3).

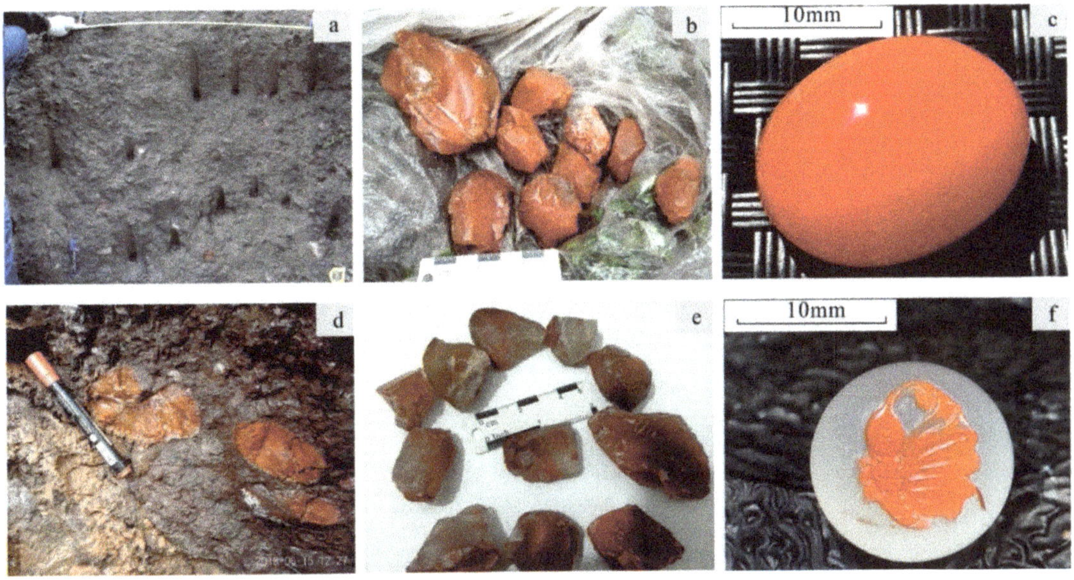

Figure 3. Field and processed photos of agate minerals. (**a**) Agate conglomerate of the Feixianguan Formation; (**b**) raw cherry-red agate; (**c**) processed cherry-red agate; (**d**) fresh cherry-red agate with red jasper outcrop; (**e**) ice-drift red agate; (**f**) processed ice-drift red agate.

3. Methodology

3.1. Field Geological Survey

A field geological survey was conducted to elucidate the ore-bearing strata of Nanhong agate, examine the sedimentary environment within the exploration area, and characterize corresponding lithological features. This process facilitated the rapid delineation of the exploration area. Initially, the survey involved the preliminary identification of lithological types, material compositions, sedimentary characteristics, ore-bearing potential, contact relationships, and spatiotemporal distribution variations in sedimentary rock layers. Concurrently, lithostratigraphic sequences were established, and an analysis of their depositional facies and environments was conducted. Furthermore, stratigraphic profiles were measured to discern ore-bearing structures and lithological assemblages. The prospecting and delineation of ore-bearing rock layers, coupled with the identification of specific lithological layers associated with mineralization, were undertaken. Additionally, structural features and mineralization alterations related to ore formation were studied. This phase of research aimed to provide a foundational understanding for subsequent geophysical exploration efforts. The locations of geological observations and sampling are marked in Figure 2.

3.2. Physical Property Measurements

The measurement of the physical properties of rock specimens contributes to the evaluation of feasibility and the interpretation of geophysical surveys [14]. Representative rock samples were collected from the field superficially and from the artificial mining cave, and they were subjected to laboratory-based measurements of magnetic and electrical resistivity properties. The samples of gravel with conglomerate agate and sandstone were taken from the cave, while the samples of basalt were collected superficially. The sampling was carried out at a few locations of the geological observation (Figure 2). Prior to laboratory measurements, samples were processed under water-cooled conditions using a tabletop drill and a cutting machine to produce cylindrical specimens with standardized dimensions: approximately 22 mm in height and 25 mm in diameter. Magnetic susceptibility parameters were determined using a British MS2 magnetic susceptibility meter [15], with a resolution of 2×10^{-6} SI units. Subsequently, following saturation in water for over 24 hours, electrical resistivity parameters were measured using an RP-1 rock and mineral electrical conductivity meter [16]. A total of 128 samples, comprising sandstone, conglomerate, and basalt, were collected for physical property measurements.

3.3. Geophysical Exploration

High-resolution magnetic survey is a geophysical method that involves observing and analyzing magnetic anomalies caused by differences in the magnetic properties of rocks and ores, thereby studying the geological structures and distribution patterns of mineral resources. It investigates the distortion of the geomagnetic field caused by the superposition of magnetic fields generated by magnetic bodies on Earth [17–21]. In this study, the presence of highly magnetic basalt may cause overlapping magnetic anomalies with conglomerates, which should be distinguished during the analysis. Total field measurements were conducted using a PMG-2 proton magnetometer manufactured by the Czech company Satisgeo [22], with diurnal variations corrected to obtain magnetic anomalies. Prior to operation, on-site performance checks were carried out for all instruments used in the study, including accuracy calculations and inter-instrument consistency tests, to ensure proper functioning and accuracy requirements.

We generally use the induced polarization method or electrical and electromagnetic methods for electrically conductive mineral exploration [9,10,23–25], and sometimes, depending on the properties of the targeted mineral and associated host rock types, we could use ground penetrating radars (GPRs) [26] and other geophysical methods. The chalcedony ore does not have polarization characteristics and has its own properties, so we used magnetic, VES, and CSAMT methods. CSAMT is an electromagnetic sounding technique that uses finite-length grounded electric dipoles as current transmitting sources to simultaneously measure electrical and magnetic fields at a certain distance from the source dipole [11,12,27,28]. It effectively delineates subsurface electrical resistivity or conductivity information, although its resolution for near-surface is relatively limited. In this study, an equatorial dipole setup was used for scalar measurements, simultaneously observing the horizontal components of the electric field (E_x) parallel to the source and the magnetic field (H_y) orthogonal to the source. Subsequently, impedance resistivity was calculated using the electric field amplitude (E_x) and magnetic field amplitude (H_y), and the impedance phase was calculated using the electric field phase and the magnetic field phase. The resistivity parameters were inverted using a joint inversion of impedance resistivity and impedance phase. The study employed the GDP32^{II}multifunctional electrical method workstation manufactured by Zonge [29], USA. The transmitter electrodes were set at AB = 800 m, receiver–transmitter separation was set at 6 ca. km (Figure 1), measuring receivers had a separation of MN = 40 m, and the frequency range was set at 8–8192 Hz, with a detection depth of 440 m.

With respect to resistivity sounding, VES employs a symmetric four-electrode configuration, in which four electrodes—A, M, N, and B(A and B as current electrodes and M and N as potential electrodes)—are arranged in a straight line and are systematically spread about the midpoint O of MN [30]. An increase in the distance between transmitting electrodes A and B is used to achieve deeper probing of the electrical characteristics of subsurface geological structures. The range of AB spans from 3 to 800 m, with a probing depth of 160 m. This method provides high resolutions for near-surface geological units but is less effective in detecting deep geological bodies and is time consuming.

Two CSAMT profiles (line 6 and line 26) were deployed along the agate mineralization zone, as shown in Figure 2. The line spacing was 200 m, and the profile direction was 150°. Finally, two resistivity sounding points were established on the profiles: VES 1 on line 6 and VES 2 on line 26. A symmetric four-electrode setup was used, with AB ranges of 3–700 m for VES 1 and 3–800 m for VES 2.

4. Results and Discussion

4.1. Geological Measurements

The Nanhong agate deposit, classified as a chalcedony-type deposit, can be further differentiated into volcanic and sedimentary types based on its genesis [31,32]. Field surveys reveal predominantly stratiform ore bodies within sedimentary rocks, manifested as conglomerates and pebbly sandstones, which are indicative of a secondary sedimentary origin. These deposits predominantly stem from the early Permian agate layers of the Emeishan basalt group. Over time, these layers underwent weathering, erosion, and redeposition, resulting in the current agate layers observed within the Feixianguan Formation. This formation, which can be subdivided into three members, is predominantly characterized by purple-red conglomerates, sandstones, and siltstones interspersed with agate-bearing conglomerate layers.

Notably, the agate-bearing layers are primarily confined to the second member of the Feixianguan Formation. These layers are characterized by medium-grained volcanic conglomerates with minor occurrences of sandstone and mudstone. These hard layers, marked by siliceous cementation, host an agate ore grade of 2–3%, predominantly featuring cherry-red agate, with some occurrences of floating ice-red and white agate. These distinctive agate-bearing layers, harder than the adjacent strata and with a gravel content ranging from 60% to 70%, are predominantly composed of volcanic rocks. The grain size within these conglomerates varies from 3 mm to 10 cm, with a predominant range of 1–5 cm. Geological observations indicate gentle dip angles of 5°–15° in the agate layers, suggesting ease of mining.

Further investigations highlight that both the roof and floor of these agate deposits pertain to the second member (T_1f^2) of the Feixianguan Formation within the Triassic system. The roof is characterized by purple-red thin–medium-grained quartz siltstones exhibiting horizontal bedding, whereas the floor is dominated by purple-red medium-grained feldspathic quartz sandstones with parallel bedding and minor cross-bedding. Specialized geological surveys pinpoint the sedimentary environment of this formation as a tidal deltaic phase and delta plain subphase. The hard-cemented agate-bearing conglomerates at the base correspond to the main river channel microfacies, while the upper sandy mudstone-cemented conglomerates are indicative of debris flow microfacies. Sandstone layers with parallel bedding represent inter-channel microfacies. The unique sedimentary environments associated with the agate layers, combined with distinct lithological features, offer valuable insights for narrowing down the exploration area, as illustrated in Figures 2 and 4.

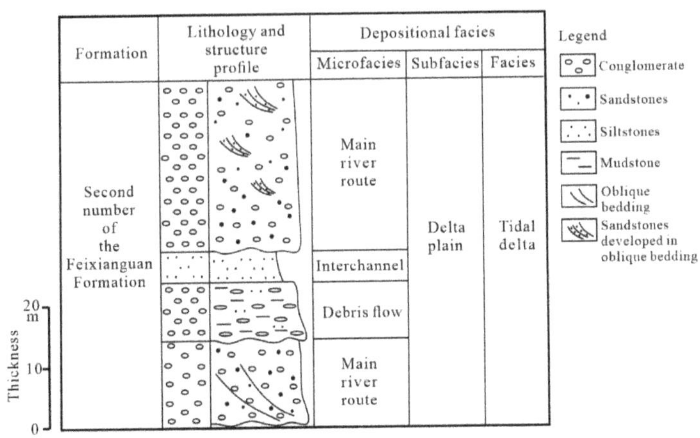

Figure 4. A representative sedimentary sequence of the second member of the Feixianguan Formation (results of this field geological survey).

4.2. Physical Property Measurements

Table 1 presents the physical property measurements of conglomerate, sandstone, and basalt samples collected from the exploration area. However, the mudstone samples in this area proved to be extremely fragile, making it impossible for us to produce suitable specimens for analysis. So, it is not shown in Table 1. The measurements reveal distinct characteristics among the rock types. Conglomerates exhibit medium to high resistivity and significant magnetic variations, generally displaying strong magnetism. In contrast, sandstones have lower resistivity and are essentially non-magnetic. Basalts demonstrate high resistivity with pronounced magnetism. Mudstones, due to their water content, exhibit low resistivity. Importantly, conglomerates containing agate display higher magnetism compared to sandstones. Thus, targeting conglomerate layers with medium characteristics can effectively guide the exploration and prospecting of agate deposits.

Table 1. Statistical parameters of rock physical properties in the study area.

Name	Samples	Susceptibility (K) ($\times 10^{-6}$ SI)			Resistivity (ρ) (Ω·m)		
		Min	Max	Mean	Min	Max	Mean
Gravel (with conglomerate and agate)	31	5059.74	18,432.87	9715.50	286.10	510.31	382.80
Sandstone	57	165.49	543.18	304.54	37.95	308.79	141.66
Basalt	40	18,493.18	78,893.85	52,644.5	957.62	2964.69	1892.01

4.3. Integrated Geophysical and Geological Section Measurements

CSAMT data were processed using the smooth-model inversion software SCSINV [33]. The RMS errors for the various measurement points ranged from 1.1 to 2.3.

Figure 5 illustrates the integrated profile of the line 6 geological and geophysical survey, sequentially presenting magnetic survey results, CSAMT data, and geological profile measurements. The topography displays notable undulations. Within the profile's central segment, magnetic anomalies remain steady, with elevated anomalies at both ends of the profile. The positions at the end of the profile indicate the presence of basalt according to the geological assessment.

The CSAMT method results delineate a 440-meter-deep section into three distinct electrical layers: a high-resistivity near-surface layer, a subsequent low-resistivity layer, and a bedrock layer characterized by high resistivity. Geological interpretations align with these electrical findings, identifying the high-resistivity layer at 30–80 m as sandstone from the Feixianguan Formation, the low-resistivity layer as mudstone from the Xuanwei Formation, and the bedrock as basalt. Notably, the sandstone of the Feixianguan Formation houses agate layers, with agate minerals discovered near mountain peaks and basalt exposures at the position of the end of the profile. The slightly higher magnetic anomalies at the position of the beginning of the profile likely indicate gravel associated with agate mineralization. However, not all moderate anomalies correspond to the agate layers. Deep-seated magnetic minerals can significantly impact surface magnetic survey results. As previously mentioned, the magnetic anomaly at the end of the profile is indicative of basalt. Consequently, vertical electrical sounding was conducted near the point 278 of line 6 to refine the depth characterization of the ore-bearing layer within the Feixianguan Formation.

Figure 6 illustrates the composite profile of the 26-line survey, which shares similarities with the 6-line results. High values of magnetic anomalies are observed at both ends of the profile. The CSAMT measurements revealed the electrical characteristics of the section, and vertical electrical sounding was carried out near the point 262 of line 26. Geological profiles also indicate the presence of agate minerals near mountain peaks and at the position at the beginning of the profile.

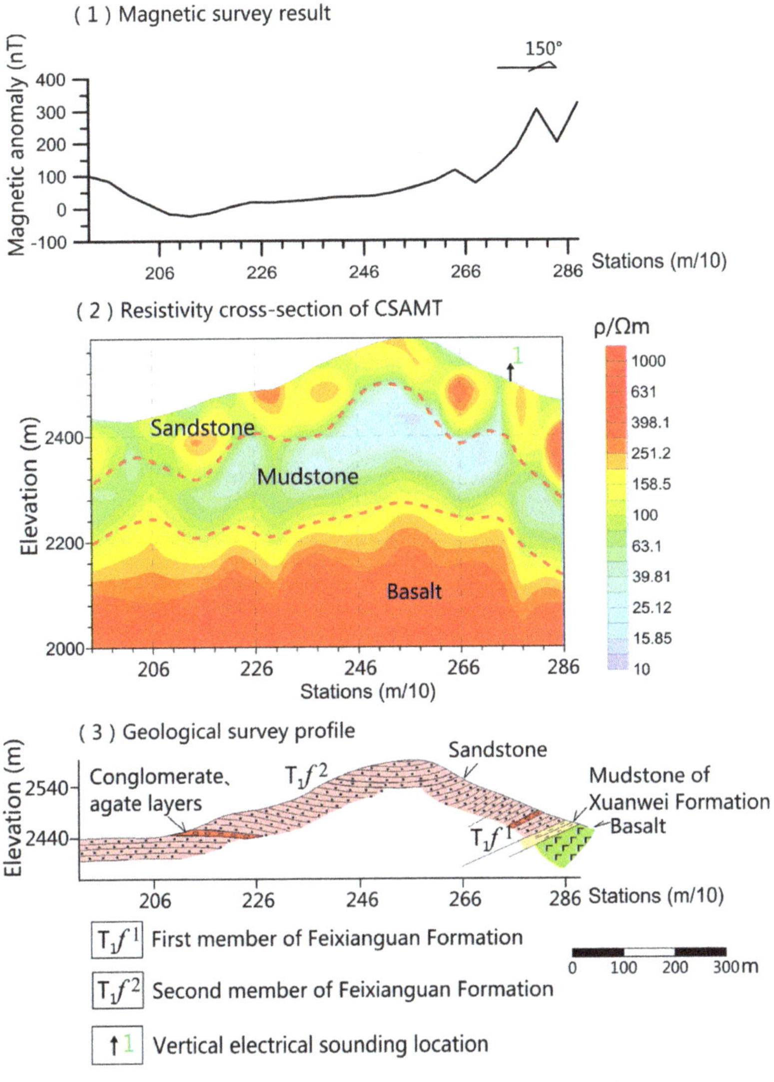

Figure 5. Geological–geophysical composite profile along the exploration line 6.

Figures 7 and 8 show the CSAMT depth soundings at various frequencies for points 206 on line 6 and 226 on line 26, respectively. This area has relatively low resistivity, with apparent resistivity values near 100 Ωm. The signal transitions to the near-field around 100 Hz. One key feature of CSAMT is that as the frequency decreases, the depth of detection increases. The curves (Figure 7) imply a simplified three-layer electrical structure from shallow to deeper depths. Formulas to convert electric and magnetic fields with respect to impedance resistivity and phases are described in [34].

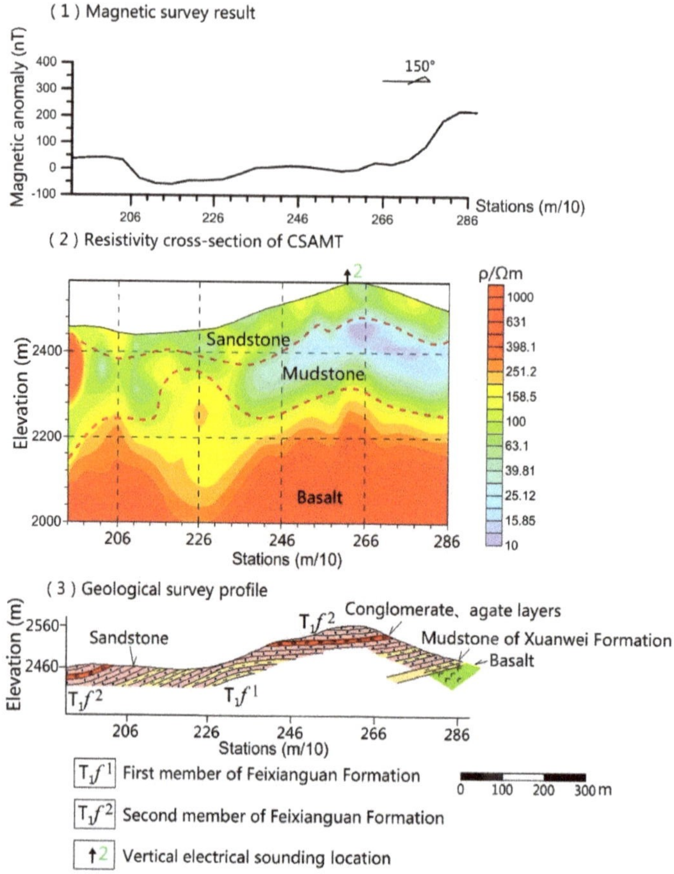

Figure 6. Geological–geophysical composite profile along exploration line 26.

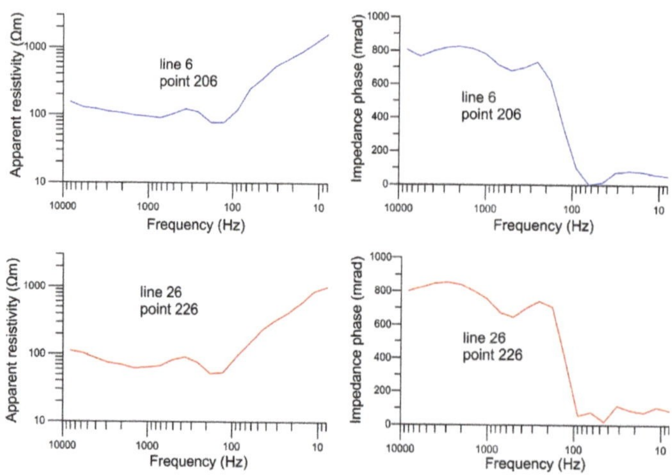

Figure 7. Controlled-source audiomagnetotelluric measurement with apparent resistivity (**left**) and impedance phase (**right**) curves.

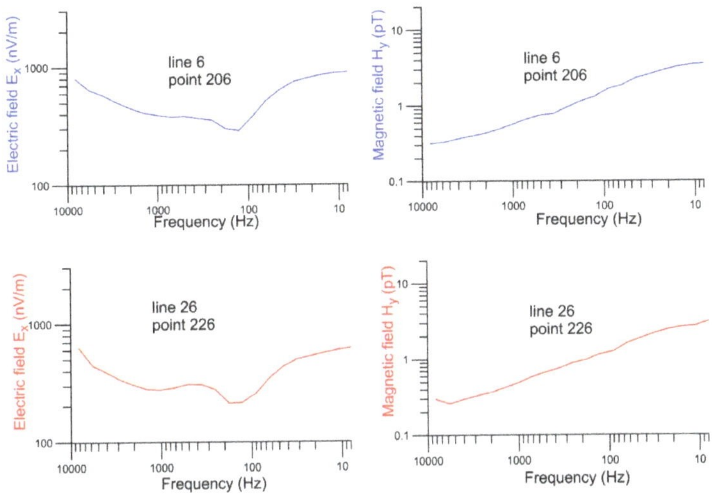

Figure 8. Electromagnetic field curves from the controlled-source audiomagnetotelluric measurement point.

According to geological observations (Figure 9), the first member (T_1f^1) of the Feixianguan Formation consists of greenish-gray lithic feldspathic sandstone, ranging in grain size from coarse to fine, with a thickness varying between 40 cm and 2 m. As one moves upwards, the lithology changes to purplish-red medium-bedded lithic feldspathic sandstone interspersed with layers of purplish-red siltstone (Figure 9a). These sandstones frequently include clasts of purplish-red mudstone and occasional calcareous nodules, while the siltstones sometimes contain small calcite crystals.

Figure 9. Characteristics of the agate-rich strata of the Feixianguan Formation. Locations of these samples are shown in Figure 2. (**a**) Interbedded sand-mudstone of the first member of the Feixianguan Formation; (**b**) agate conglomerate layer of the second member of the Feixianguan Formation; (**c**) wedge-shaped cross-bedding in the second member of the Feixianguan Formation; (**d**) mudcracks in the third member of the Feixianguan Formation.

The second member (T_1f^2) of the Feixianguan Formation is composed of purplish-red medium-bedded lithic feldspathic sandstone interlayered with purplish-red agate-bearing lithic feldspathic sandstone, siltstone, or mudstone (Figure 9b). This section is notable for its abundant agate clasts within the sandstone, which exhibits parallel bedding, tabular cross-bedding, and wedge-shaped cross-bedding (Figure 9c).

Vertical electrical sounding (Figure 10) offers high-resolution insights into the subsurface. We used the IP12win software from Moscow State University for the interpretation. The RMS error of point 01 and point 02 is, respectively, 1.2 and 1.6. It showed good data fitting in Figure 11. The results reveal a near-surface low-resistivity layer corresponding to water-rich sandstone, overlying the higher-resistivity gravel-bearing sandstone of the Feixianguan Formation. Beneath this, the sequence transitions to the Xuanwei Formation mudstone and bedrock. Notably, from VES 1 to VES 2, the burial depth of the agate layer progressively decreases.

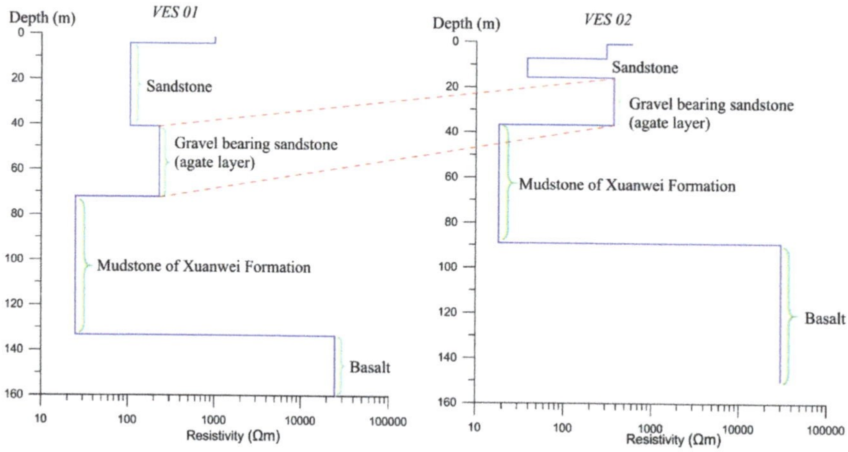

Figure 10. Inversion results of the vertical electrical sounding points.

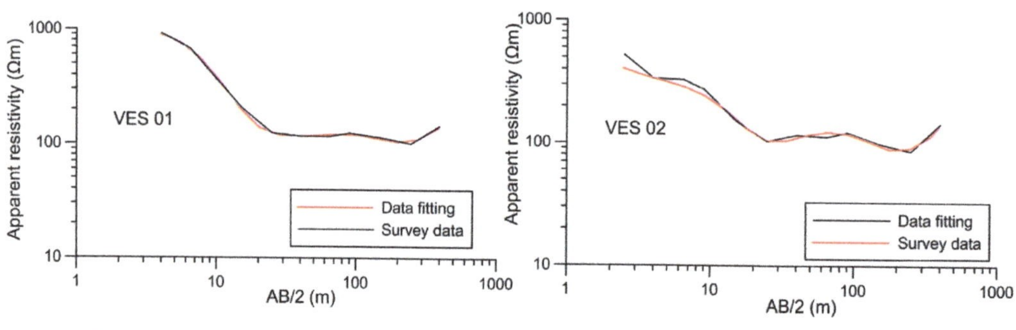

Figure 11. Data fitting for vertical electrical sounding.

4.4. Practicality of Integrated Geological and Geophysical Measurements in Agate Exploration

The formation of Nanhong agate in the Liangshan region of Sichuan, China, is primarily influenced by volcanic and sedimentary processes. Late Permian tectonic stresses, mainly manifesting as east–west extensional forces, resulted in the development of north–south trending extensional faults, which triggered extensive volcanic activity. These faults served as conduits for the extrusion and shallow intrusion of basaltic magma and subsequently facilitated the deposition of Nanhong agate through the filling of early volcanic rock cavities by post-volcanism hydrothermal fluids. Additionally, the transformation of primary agate through weathering, erosion, transportation, and diagenesis contributes to the formation of agate deposits. Sedimentary processes also play a crucial role in this formation.

Our exploration of this distinctive agate-type quartz mineral commenced with geological surveys. By analyzing the unique sedimentary environment and petrological characteristics of the exploration area, we identified geological formations and rock compositions. These foundational data facilitated the rapid delineation of potential mineralized sedimentary strata. Our findings revealed that agate deposits predominantly occur as stratiform or stratiform-like bodies within sedimentary rocks, often appearing as conglomerates or pebbly sandstones, classified as secondary sedimentary deposits. The second conglomerate layer of the Feixianguan Formation is the primary ore-bearing horizon for agate. It exhibits a gentle dip ranging from 5° to 12°, aligning closely with the terrain due to its mild inclination. The grain size is predominantly coarse, consisting mainly of purplish-red fine–medium-grained sandstone interbedded with purplish-red mudstone containing agate. Agate pebbles are commonly found within the sandstone, indicating significant sedimentary hydrodynamics during deposition. Sedimentary structures such as parallel bedding, tabular cross-bedding, and wedge cross-bedding are highly developed, reflecting the strong sedimentary hydrodynamics of the Feixianguan Formation's second conglomerate layer. This layer's ore formation is primarily attributed to an earlier formed agate layer of the Emeishan basaltic formation during the late Permian, which underwent weathering, erosion, transportation, and re-deposition, culminating in the current conglomerate agate layer of the second member of the Feixianguan Formation. The sedimentary environment is predominantly characterized by alluvial fan or deltaic facies. The magmatic rocks in the mining area mainly belong to the Emeishan basaltic group. While ore-controlling structures are not prominently visible and alteration is minimal, the agate primarily exhibits a single-layer structure with colors mainly in shades of cherry red, and they exhibit ice-floating patterns (Figure 3).

Upon delineating the preliminary agate deposit zones through basic field geological surveys, we utilized physical property measurements for geophysical interpretation. High-precision magnetic surveys were conducted to identify potential agate-bearing areas. The CSAMT method was employed to probe deep electrical structure, thereby determining the thickness of the ore-bearing strata and the depth to bedrock. Further precision in determining the depth and thickness of the agate layer was achieved through vertical electrical sounding. Table 2 presents the comparison results between CSAMT, resistivity sounding, and physical property measurements. As shown in the table, the resistivity ranges of geological units measured by various geophysical methods are generally consistent with those obtained from physical property measurements.

Table 2. Comparison of survey results of geophysical methods and resistivity measurements for different geological units.

	Result of CSAMT	Result of Resistivity Sounding	Resistivity Measurements
Gravel bearing agate and conglomerate	100–400 Ωm	200–400 Ωm	382.8 Ωm
Sandstone	100–400 Ωm	20–100 Ωm	141.66 Ωm
Mudstone	10–100 Ωm	20–30 Ωm	/
Basalt	>400 Ωm	>20,000 Ωm	1892.01 Ωm

Previous studies have recognized the shallow burial depth and non-polarizing characteristics of agate, employing ground-penetrating radar for prospecting. Measurements were taken at several study test sites with respect to the subsurface geology of weathered melaphyre and pyroclastic deposits using a GPR system (ProEx) [26]. However, no precedent exists for agate exploration using the combination of geological and electrical methods employed in this study. Therefore, our study integrates geological surveys with high-precision magnetic surveys, CSAMT, and vertical electrical sounding to construct an agate ore prediction map (Figure 12). The predictive map was constructed using the stratigraphic sequence obtained from geological surveys, incorporating the medium magnetic anomaly of the agate gravel layer and the resistivity parameter values measured for various geological units based on their physical properties, as shown in Table 2.

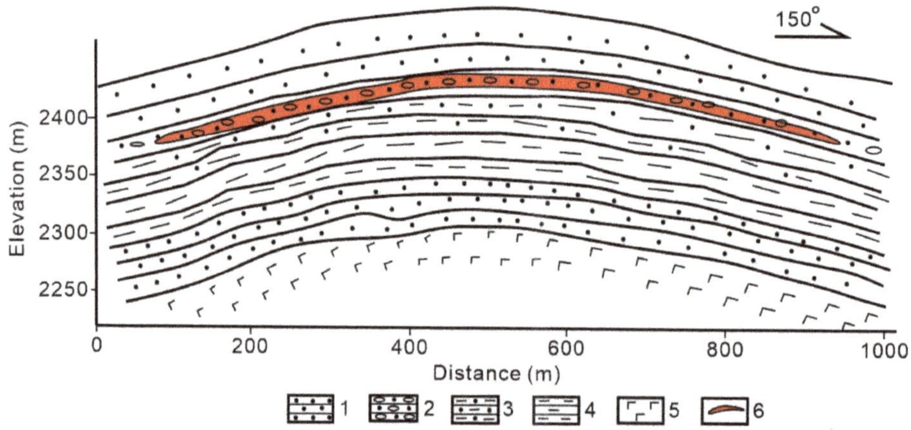

Figure 12. Predicted mineral body map of the agate deposit in the exploration area. 1—Sandstone; 2—conglomeratic sandstone; 3—silty sandstone; 4—mudstone; 5—basalt; 6—agate. Location of this predicted mineral is shown in Figure 2.

Our extensive survey indicates that the surface is largely dominated by sandstone, with agate deposits found within the conglomerate layer of the second member of the Feixianguan Formation, which comprises pebbly sandstones. This layer is underlain by the mudstones of the Xuanwei Formation and sits atop the Emeishan basalt bedrock. According to the ore prediction map, agate deposits are identifiable by targeting high-resistivity and magnetically active conglomerate layers within the second member of the Feixianguan Formation. Consequently, the applied methods and techniques for exploring agate-bearing horizons have proven effective. This study confirms that geological surveys, physical property measurements, and geophysical methods, particularly high-precision magnetic surveys and CSAMT, are powerful tools for chalcedony mineral exploration.

5. Conclusions

In this study, geological investigations were undertaken to identify the ore-bearing strata, sedimentary environments, and petrological characteristics of agate deposits in the Liangshan region of Sichuan, China. This strategy significantly narrowed the exploration area. Advanced geophysical exploration methods, including high-precision magnetic surveys, CSAMT techniques, and vertical electrical sounding, were then applied to enhance differentiation capabilities. These techniques successfully identified the agate-bearing strata, the low-resistivity mudstones of the Xuanwei Formation, and the underlying bedrock, leading to the creation of a comprehensive mineral prediction map that details the distribution of ore-bearing agate layers.

Our results indicate that the agate deposits are primarily located within the conglomeratic sandstone layer of the second member of the Triassic Feixianguan Formation, characterized by relatively high resistivity and a thickness of approximately 40 m. Beneath this layer lies the low-resistivity mudstones of the Xuanwei Formation, underlain by the highly magnetic Permian Emeishan basalt.

This research provides a solid scientific basis for the precise identification and evaluation of Nanhong agate deposits and introduces new perspectives and methodologies for exploring similar quartz mineral deposits. We believe that the insights gained from this study will significantly contribute to the responsible exploitation and utilization of Nanhong agate resources. Additionally, the methods and findings presented are expected to serve as a crucial reference for geophysical exploration efforts targeting quartz mineral deposits worldwide.

Author Contributions: Conceptualization, methodology, soft, formal analysis, writing: S.G. and J.Y.; geological investigation: S.Y. and K.Z.; geophysical investigation, geophysical data methodology and analysis: S.G.; Project manager: Y.L.; Document editing: J.Y. and M.W. All authors have read and agreed to the published version of the manuscript.

Funding: This research was funded by the China Geological Survey Program [grant number: DD 2024-2247, DD20230309, DD20190032, and DD20160313-24]; the National Key R&D Program of China [grant number: 2018YFE0208300].

Data Availability Statement: Data are available to request from the China Geological Survey. https://geocloud.cgs.gov.cn/.

Acknowledgments: Please allow us to show our great appreciation to the editor and reviewers for their patience and helpful comments and suggestions.

Conflicts of Interest: The authors declare no conflicts of interest.

References

1. Huang, Y.F.; Hua, D.J.; Jia, Q. Characteristics and prospecting prospects of agate deposits in the Lomoyida area of Sichuan Province. *China Sci. Technol. Inf.* **2022**, *8*, 73–75.
2. Zhou, Y.; Liu, Z.; Zhao, Z.; Guo, Y. Quantitative study on colour and spectral characteristics of Beihong agate. *Minerals* **2022**, *12*, 677. [CrossRef]
3. What Is South Red Agate in Chinese Jewelry? 2023. Available online: https://www.chinese-showcase.com/blogs/jewelry/what-is-south-red-agate-in-chinese-jewelry (accessed on 2 August 2023).
4. Tang, F.; Lin, F.; Jin, C.; Deng, G.; Zhao, H.; Tie, Y.; Yu, Q.; Ren, S.; Ren, G. Practical geological technologies facilitating poverty alleviation in Wumeng Mountain area. *China Geol.* **2020**, *3*, 504–508. [CrossRef]
5. Chen, S.Y.; Wang, S.Q.; He, X.M.; Luo, B.; Zhou, C.J.; Shi, X.P. Preliminary study of the characteristics and genesis of the Nanhong agate deposit in Liangshan, Sichuan, China. *Jewel. Technol.* **2015**, *1*, 196–199.
6. Jing, H.X.; He, Y.G.; Luo, B.; Li, J.J.; Liu, W.; Wang, N. Characteristics of occurrence and genesis of Nanhong agate deposit in Waxi area, Meigu county, Sichuan province. *Contrib. Geol. Miner. Resour. Res.* **2023**, *38*, 326–330.
7. Baranwal, V.C.; Rønning, J.S.; Larsen, B.E.; Su, Y.; Zhang, B.; Liu, Y.; Ren, X.; Gautneb, H.; Gellein, J. A case history of graphite exploration in north Norway integrating various geophysical surveys. *Minerals* **2024**, *14*, 266. [CrossRef]
8. Gong, S.; Yang, Y.; Lin, P.; Wu, W.; Zheng, C.; Shi, F.; Wu, X.; Weng, A.; Zhang, G.; Gu, G.; et al. Three-dimensional electrical exploration methods for the mapping of polymetallic targets in Gansu Province, China. *Geophys. Prospect.* **2019**, *67*, 1929–1947. [CrossRef]
9. Gong, S.P.; Yang, Y.B.; Zhang, G.Z.; Lin, P.R.; Wu, W.L.; Zhen, C.J.; Shi, F.S. Experimental study on 3D time-domain electromagnetic exploration in Huaniushan Pd-Zn deposit. *Sci. Technol. Eng.* **2018**, *18*, 16–24.
10. Gong, S.P.; Lu, G.F.; Xi, M.J.; Ma, S.M.; Su, W.L. The application of integrated geophysical and geochemical methods to the prospecting of copper polymetallic deposits in the arid desert area. *Geophys. Geochem. Explor.* **2021**, *45*, 1–10.
11. Yang, J.; Zhao, L.; Shen, Y.; Shi, K. Application research of controlled source acoustic magnetotelluric in the location forecasting of buried orebodies. *Geol. Sci. Technol. Inf.* **2000**, *19*, 107–112.
12. Zhang, G.; Li, R. The test result of the controlled audio-frequency magnetotelluric method in the prospecting for deep ore deposits. *Geophys. Geochem. Explor.* **2010**, *34*, 66–70.
13. Huang, H.; Cawood, P.A.; Hou, M.; Yang, J.; Ni, S.; Du, Y.; Yan, Z.; Wang, J. Silicic ash beds bracket Emeishan Large Igneous province to <1 m.y. at ~260 Ma. *Lithos* **2016**, *264*, 17–27.
14. Tschirhart, V.; Morris, B. Physical Rock Property Constraints and Regional Geophysical Interpretations of the Northeast Thelon Basin, Nunavut. 2017. Available online: https://www.academia.edu/70174978/Physical_Rock_Property_Constraints_and_Regional_Geophysical_Interpretations_of_the_Northeast_Thelon_Basin_Nunavut (accessed on 25 April 2024).
15. Kozhevnikov, N.O.; Kamnev, Y.K.; Kazansky, A.Y. Error analysis of frequency-dependent magnetic susceptibility measurements: Magnetic viscosity studies with the Bartington MS2 system. *Russ. Geol. Geophys.* **2014**, *55*, 508–514. [CrossRef]
16. Li, L.; Guo, Y.; Chen, X. Development of RP-1 type rock and ore electrical measurement system. *Geophys. Geochem. Explor.* **2013**, *3*, 529–532.
17. Hallimond, A.F.; Herroun, E.F. Laboratory determinations of the magnetic properties of certain igneous rocks. *Proc. R. Soc. London. Ser. A Contain. Pap. A Math. Phys. Character* **1933**, *141*, 302–314.

18. Legowo, B.; Irfan, N.M.; Suryanto, W.; Purnama, B. Study comparison physical properties and minerals analysis of rock magnetic in Lawu Volcanic Complex. *J. Phys. Conf. Ser.* **2023**, *2498*, 012033. [CrossRef]
19. Essaa, K.S.; Nadyb, A.G.; Mostafab, M.; Elhussein, M.S. Implementation of potential field data to depict the structural lineaments of the Sinai Peninsula, Egypt. *J. Afr. Earth Sci.* **2018**, *147*, 43–53. [CrossRef]
20. Biswas, A.; Rao, K. Interpretation of magnetic anomalies over 2D fault and Sheet-Type mineralized structures using very fast simulated annealing global optimization: An understanding of uncertainty and geological implications. *Lithosphere* **2021**, *2021*, 2964057. [CrossRef]
21. Essa, K.S.; Elhussein, M. Magnetic interpretation utilizing a new inverse algorithm for assessing the parameters of buried inclined dike-like geological Structure. *Acta Geophys.* **2019**, *67*, 533–544. [CrossRef]
22. Petronis, M.; Valenta, J.; Rapprich, V.; Lindline, J.; Heizler, M.; Vries, B.; Shields, S.; Balek, J.; Fojtíková, L.; Tábořík, P. Emplacement history of the Miocene Zebín Tuff Cone (Czech Republic) revealed from ground geophysics, anisotropy of magnetic susceptibility, paleomagnetic, and 40Ar/39Ar geochronology data. *Geochem. Geophys. Geosystems* **2018**, *19*, 3764–3792. [CrossRef]
23. Amaral, P.A.C.; Borges, W.R.; Toledo, C.L.B.; Silva, A.M.; Godoy, H.V.; Santos, M.H. Electrical prospecting of gold mineralization in exhalites of the Digo-Digo VMS occurrence, central Brazil. *Minerals* **2023**, *13*, 1483. [CrossRef]
24. Nazari, S.; Rochlitz, R.; Günther, T. Optimizing semi-airborne electromagnetic survey design for mineral exploration. *Minerals* **2023**, *13*, 796. [CrossRef]
25. Wu, Y.; Xie, H.; Ji, Y.; Zhao, P.; Wang, Y. The potential of the horizontal component TEM data in the detection of polarizable mineral: Synthetic cases. *Minerals* **2023**, *13*, 523. [CrossRef]
26. Mierczak, M.; Karczewski, J. Location of agate geodes in Permian deposits of Simota gully using the GPR. *Acta Geophys.* **2021**, *69*, 655–664. [CrossRef]
27. Sandberg, S.K.; Hohmann, G.W. Controlled-Source Audiomagnetotellurics in Geothermal Exploration. Topical Report. *Geophysics* **1980**, *47*, 100–116. [CrossRef]
28. Wang, J.; Wang, M.; Liu, W.; Lin, P. The full-field apparent resistivity of CSAMT defined by the magnetic field y-component and an application to a geological survey in the Xiong'an New Area, China. *Bull. Geophys. Oceanogr.* **2022**, *63*, 311–324.
29. Atya, M.; Meneisy, A.M.; Bedair, S. Integration of electrical resistivity and time-domain electromagnetic to investigate the subsurface stratigraphic conditions at Assiut Power Plant, Assiut, Egypt. *J. Astron. Geophysics.* **2020**, *9*, 471–482. [CrossRef]
30. Hamzah, U.; Samsudin, A.R.; Malim, E.P. Ground water investigation in Kuala Selangor using vertical electrical sounding (VES) surveys. *Environ. Geol.* **2007**, *51*, 1349–1359. [CrossRef]
31. Shen, M.; Lu, Z.; He, X. Mineralogical and geochemical characteristics of banded agates from Placer deposits: Implications for agate genesis. *Acs Omega* **2022**, *7*, 23858–23864. [CrossRef]
32. Yan, S.; Zhao, C.; Qin, Y.; Bai, X. The geological characteristics and mineralization regularities of the Nanhong agate deposit in Zhaojue-Meigu area. In Proceedings of the 1st China Exploration GeoScience Conference (CEGC), Hefei, China, 15–20 September 2021.
33. Macinnes, S.; Raymoned, M. *SCSINV Documentation Zonge Data Processing Smooth-Model CSAMT Inversion Version 2.20*; Zonge Engineering and Research Organization, Inc.: Tucson, AZ, USA, 2006.
34. *DZ/T 0280-2015*; Technical Regulation for Controlled Source Audio Magnetelluric Method. Geological Press: Beijing, China, 2015.

minerals

MDPI

Review

Case Studies of Magnetic and Electromagnetic Techniques Covering the Last Fifteen Years

Marc A. Vallée *, Mouhamed Moussaoui and Khorram Khan

Geo Data Solutions GDS Inc., 4402 Louis-B. Mayer, Laval, QC H7P 0G1, Canada
* Correspondence: mvallee@geodatasolutions.ca

Abstract: Magnetic and electromagnetic techniques have a long history of application in mineral exploration to detect deposits and their surroundings. Their implementation over the last fifteen years has been affected by strong variations in the mining market in parallel with important technological developments. During this period, both methods were the subject of numerous documented case studies all over the globe, which is a sign of popularity and longevity of these techniques. Through a review of case histories from the main geophysical journals, we analyze the principal usage of these methods when applied to mineral exploration, while the majority of documented cases originate from North America, Asia, and Australia. There are more case studies describing the use of the magnetic method and we attribute this popularity to direct and indirect use of this method for mineral exploration. In particular, there is an increasing number of magnetic surveys conducted with drones. Combining magnetic and electromagnetic techniques is also common. The number of magnetic and EM technique case histories range by descending order from gold, porphyry copper, polymetallic, massive sulfides, uranium, Ni-Cu-PGE, iron ore, kimberlite, and iron-oxide copper-gold, with a number of single continent-specific applications.

Keywords: magnetic exploration; electromagnetic exploration; case studies

1. Introduction

Magnetic techniques are mainly based on the measurement of the Earth's magnetic field [1] over a given area and the associated geographical variations with changes in the magnetic susceptibility of the subsurface [2]. Electromagnetic methods (EM) [3] use natural or artificial, electric or magnetic sources to generate secondary signals due to the presence of conductivity/resistivity subsurface variations [4]. Both techniques, either ground, borehole, or airborne, are widely used in mineral exploration [1,5] either to directly detect mineralization or to characterize the geology surrounding a potential deposit. The last fifteen years were an active period for the development of these techniques. Years 2010–2012 saw a significant allocation for exploration budgets [6]. However, these years were followed by four years of decreasing exploration budgets that slowly recovered after 2017 without reaching their previous peak. Worldwide production reached a plateau around 2013 [7] and has been stable since. During the same period, technology improvements have had an profound impact on society [8,9], and in particular exploration geophysics.

Technological advances in the application and interpretation of these techniques have been numerous over the last fifteen years, since the seminal works of [1,5,10,11] covering the first decade of the XXIst century, and are difficult to relate succinctly. For this reason, we refer the reader to the reviews of [12–20] which cover acquisition and interpretation. Thus, the focus of this paper is on case studies of magnetic and EM techniques, since they are an important part of the geophysical literature that reflects technological advances and their usage. We mainly rely on papers published in the last fifteen years in *Geophysics*, *Exploration Geophysics*, *Geophysical Prospecting*, *Minerals*, and *Australian Society of Exploration Geophysicists (ASEG) Extended Abstracts*, as these publications can be easily found on the internet. For developers, these publications are opportunities to justify claimstaking while also celebrating success sometimes. Consequently, case studies published since 2010 highlight the achievements of the last fifteen years. First, we start with applications of aeromagnetic drone techniques, followed by publications of magnetic and EM techniques, classified by continent, knowing that our choice of publications may bias the geographical distribution of our results. Magnetic surveys are often conducted with gravity surveys and in this circumstance classified as potential field surveys.

2. Drones

During the period covered, various studies were conducted to compare unmanned airborne vehicle (UAV) measurements with ground or airborne measurements. Ref. [21] conducted a simulated magnetic UAV survey and compared the results with ground and airborne surveys. The results have higher resolution than airborne surveys and are similar to ground survey data. Ref. [22] compares the responses of UAV magnetic data with airborne data over a hilly area in China. The results compare favorably. Ref. [23] compares measurements collected with a rotary-wing UAV magnetometer and with fixed-wing aeromagnetic data over iron-oxide deposits in central Sweden. The two data sets outline the mineralization similarly. Ref. [24] conducted an UAV survey over mineralization in New Brunswick, Canada, whose results compare favorably with upward continued ground magnetic data. Ref. [25] compares the response of ground, helicopter and UAV magnetic data over a prospective gold area in northern Québec. UAV data are of the same quality as traditional methods. Ref. [26] conducts a combined ground and UAV magnetic survey in northern Sweden, where the ground and airborne surveys were compared. Strong correlation is observed between the two surveys. Ref. [27] describes trials of an UAV sub-audio magnetics (SAM) survey at Forrestania Electromagnetic Test Range in Western Australia.

Several drone case studies were conducted over known mineralizations. Ref. [28] presents the results of a magnetic UAV survey conducted in the Shebandowan Greenstone Belt, northwest of Thunder Bay, Ontario, Canada. Ref. [29] describes a magnetic UAV survey over a gold property in northern Québec, Canada, and after repeatability tests for validating the data, these data were inverted with Geosoft VOXI toolbox. Ref. [30] presents a case study of helimagnetic survey for regional exploration and UAV for high-resolution exploration of iron ore located in Pocheon, Korea. Ref. [31] developed a new iterative imaging technique applied to UAV magnetic data for achieving high resolution and identifying weak anomalies to be applied over an iron deposit in China. Other UAV studies include a survey of the Bjerkreim–Sokndal layered intrusion in southwestern Norway with custom multirotor magnetic UAV, in an area with challenging environmental conditions [32]. The UAV added information about the important geological contacts and, with multiple flight altitudes, helped in the construction of the anomaly. Ref. [33] describes a combined ground–UAV TDEM survey carried out over a gold prospect in eastern Russia. Ref. [34] presents the results of an UAV magnetic gradiometry survey over a little iron ore deposit in western Iran.

3. Magnetics

3.1. *Africa*

Ref. [35] presents a potential field study of kimberlites in Botswana, where geometrical properties of two of the pipes were estimated using gravity and magnetic analysis, and modelling. Ref. [36] applies digital edge detector operations on magnetic data to enhance the delineation and interpretation of geological features within the Middle Benue Trough of northern Nigeria. Ref. [37] applies a joint inversion to combined gravity and magnetic datasets collected over kimberlite pipes, also in Botswana. Ref. [38] combines aeromagnetic data and geochemical analysis to map related gold ore mineralization deposits around the Wadi El-Saqia area in the Central Eastern Desert of Egypt. Ref. [39] presents magnetic data processing and interpretation for exploration of rutile from the Minta area, in Haute-Sanaga,

Cameroon. Using magnetic data, Ref. [40] establishes connections between geological structures and precious and base metal deposits of the Pangar-Djérem Zone, Cameroon. Ref. [41] analyzes aeromagnetic data, transformations, and geological information for the structural interpretation of the location of precious stones from the mineral-rich zones in parts of Lafiagi and Pategi areas of Bida basin, central Nigeria. Ref. [42] integrates ground geophysical methods (ground magnetics, electrical resistivity, and induced polarization) in conjunction with fire assay and inductively coupled plasma-atomic emission spectrometry techniques to delineate orogenic gold mineralization potential zones in the Kushaka greenschist belt, Nigeria. Using ground magnetic data, Ref. [43] characterizes anomalous bodies in the manganese-rich mining zone situated in Bouarfa, Morocco. Ref. [44] analyzes geological and geophysical data to assess the structural trends affecting phosphate distribution near Wadi El-Nakheel, Egypt. Ref. [45] presents an integrated geophysical investigation for potential gold mineralized zones located in northwestern Nigeria. Ref. [46] applies surface geometry inversion to a magnetic dataset acquired over two kimberlite pipes located in north-central Botswana. Ref. [47] describes ground and airborne geophysical data collected over a gold deposit located in Guinea.

3.2. *Asia*

Using gravity and magnetic data, Ref. [48] develops an integrated geological model for the polymetallic Shizishan ore field in Anhui Province of China. Ref. [49] presents an interpretation of aeromagnetic data collected over a north-east trending orogenic gold belt of the Piranshahr-Sardasht-Saqqez Zone, Iran. Ref. [50] presents the geophysical signatures of the Beldih mine in eastern India, known for Nb–rare-earth-element–uranium mineralization. Ref. [51] proposes a combination of magnetic data filters to locate porphyry copper deposits in the tertiary magmatic belts of Iran. Ref. [52] applies constrained inversion to the interpretation of the Macheng iron deposit in China. Ref. [53] presents the case of a large magnetic anomaly detected by airborne geophysics at Gadarwara Madhya Pradesh, in north-central India. Drilling shows that this anomaly, which can be fitted with 2D models, is caused by an iron formation. Ref. [54] applies factorial kriging to regional–residual separation of magnetic data collected over the Bashmaq area, an Iron Oxide Copper-Gold (IOCG) deposit, in northwestern Iran. Ref. [55] integrates aeromagnetic, radiometric, and satellite imagery data over the Chahargonbad area in the Kerman province of Iran, prone to Cu-bearing mineralization. Ref. [56] discusses aeromagnetic exploration methods applied to exploration of iron deposits located in the Hubei province of China. Ref. [57] applies a methodology to analyze and fuse information (Landsat, aeromagnetic data, geological data, and geochemical stream sediment analysis) over an area known for copper indices and mines in Iran. Ref. [58] presents a study of ground magnetic and electrical resistivity/chargeability data from Kerala, India, known for vein-type gold mineralization. Ref. [59] employs 3D magnetic susceptibility inversion of aeromagnetic data over lead-zinc polymetallic deposits in Yichun, China. Ref. [60] interprets magnetic anomalies of the Galinge iron polymetallic deposit in western China.

3.3. *Australia*

Ref. [61] presents the results of an aeromagnetic and radiometry survey over the Carrapateena iron-oxide copper-gold deposit located in South Australia. Ref. [62] describes various magnetic data collection and processing over the magnetite iron ore resource Hawsons Iron project in western New South Wales. Ref. [63] applies the method developed by [64] to analyze magnetic data over a gold-mineralized system in Queensland. Ref. [65] studies the anisotropy of the Monakoff carbonate-hosted IOCG in Queensland and surrounding banded iron formations to differentiate the structural and metallogenic controls. Ref. [66] compares magnetic responses from an iron-rich gossan in a volcanic environment and a limestone-hosted manganese deposit of Papua New Guinea. Ref. [67] applies geological constraints to the inversion of magnetic data over the Darlot-Centenary Gold Mine of Western Australia. Ref. [68] presents airborne magnetic inversions over the

Wallaby gold deposit of Western Australia. Ref. [69] applies petrology and petrophysics to the interpretation of airborne magnetic surveys conducted over the Mt Leahy Tenement porphyry copper project in Papua New Guinea. Ref. [70] integrates drill core geochemistry, spectral and petrophysical logs, and geophysical data to characterize the Punt Hill Copper-Gold prospect and IOCG system of South Australia. Ref. [71] interprets magnetic, radiometric, and gravity data from the Stanthorpe region, Australia, with a particular interest on hydrothermal alteration. Ref. [72] presents interpretation of potential field data over the Darlot gold mine located in the Yilgarn Craton. Ref. [73] studies the region of Heazlewood-Luina-Waratah, a prospective region for minerals in northwestern Tasmania, Australia, by combining the results of potential field (geometry and property) inversion, petrophysical measurements, and updated field mapping. Ref. [74] presents an integration modelling and associated machine learning targeting of the Jaguar Massive Sulfide (MS) deposit of Western Australia using potential field and geological information.

3.4. Europe

Ref. [75] analyzes the results of magnetic and electric surveys completed to extend known Fe-rich emery horizons and to locate new deposits in the Elmacik area, Yatagan, Turkey. Ref. [76] develops and applies a joint inversion algorithm on 3D potential field data collected over a gabbro intrusion in northern Sweden. Ref. [77] investigates, using potential fields, Slingram moving loop data and rock's physical properties the geological geometry in the Gällivare mining area, Sweden.

3.5. North America

Ref. [78] inverts magnetic and gravity data with geological and geophysical constraints over the Rambler Rhyolite in Newfoundland, Canada. Ref. [79] applies an automatic lineament network extraction method to identify magnetic lows that may represent faults from Wopmay Orogen in northwestern Canada, an area with promising polymetallic hydrothermal mineral occurrences. Ref. [80] derives constraints from a geophysical–geological feedback process and applies these constraints on inversion models for interpreting the northeast Amer belt located in Nunavut, Canada. Ref. [81] uses the normalized source strength for interpretation of potential field tensor data collected for exploration of nickel, copper and platinum group element (Ni-Cu-PGE) deposits in the McFaulds Lake area, northern Ontario, Canada. Ref. [82] analyzes aeromagnetic data to locate intrusives near the Pebble deposit in Alaska, USA. In Canada, the Sudbury structure in Ontario, which is host to several ore deposits, has the been the subject of various studies. Ref. [83] models airborne gravity and magnetic profiles, while constraining the results based on seismic sections, geological contacts, and petrophysical data of the Sudbury area. Ref. [84] predicts the 3D geological setting of the Sudbury structure by integrating available geophysical and geological information using 3D GeoModeller software. Ref. [85] builds a litho-prediction model from potential field data-constrained inversions over the Victoria property, in Sudbury. Ref. [86] analyzes potential field data covering the southwest Thelon Basin, Northwest Territories, Canada, a prospective area for uranium and other economic metals. Ref. [87] applies magnetization inversion at the regional scale and cooperative inversion of DC/IP and magnetics at the local scale over the Newton Au-Ag deposit in British Columbia, Canada. Ref. [88] compares unconstrained and constrained inversions applied on potential field data collected from three different mining sites in Canada. Constrained inversion provided more meaningful results in the three cases. Ref. [89] compares the results of three different inversions programs applied on the same dataset, Highland Valley copper district in British Columbia, Canada, with unconstrained and constrained models. Variations in the results are attributed to the intrinsic errors in producing images from different programs. Ref. [90] proposes, based on 3D magnetic data inversion, computation of an alteration index to estimate porphyry copper alteration at depth at Highland Valley. Ref. [91] presents a case study of potential field joint inversion collected over the McArthur River area of the eastern Athabasca Basin in western Canada. Airborne and ground gravity and magnetic data over one of the Tli Kwi Cho kimberlites in Northwest Territories, Canada, were interpreted with petrophysically and geologically-guided joint inversions by [92]. Ref. [93] images oxide-apatite deposits located in eastern Adirondack Highlands, upstate New York, USA, using aeromagnetic, aeroradiometric, ground gravity, petrophysical and geochemical data. Ref. [94] applies self-organizing maps (SOMs) to magnetic, radiometric and gravity data sets from the Baie Verte Peninsula, in Newfoundland, Canada. Ref. [95] incorporated 3D magnetic inversion in the structural study of Sixtymile Gold district, Yukon, Canada. Ref. [96] performs joint potential field inversion and geology differentiation to obtain a quasi-geological model of the Elk Creek Carbonatite complex, southeast Nebraska, USA, an area hosting niobium and rare earth element mineralization.

3.6. South America

Ref. [97] uses the SOM approach to analyze airborne geophysical data collected over the Brazilian Amazon. Ref. [98] analyzes aeromagnetic data collected over the Morro do Leme nickel deposit, Brazil, with an initial model based on magnetic source parameter estimation techniques. Ref. [99] presents a study characterizing geology from inverted potential field data over an iron formation in Minas Gerais, Brazil. Ref. [100] inverts magnetic data above the Furnas southeast IOCG deposit at a low magnetic latitude in Carajás Mineral Province, Brazil. Ref. [101] analyzes the geophysical data above the Catalão I alkaline–carbonatite complex, located in central Brazil, which is one of the main producers of niobium and phosphates in the world. Ref. [102] studies and delineates, through various geophysical parameters, the source of a magnetic anomaly near the Buracao da Velha copper deposit in Brazil. Ref. [103] develops an interpretation scheme based on geophysical models obtained from 2D and 3D inversion of geophysical and sparse geological data. This scheme was applied to the interpretation of geophysical data collected over the Cristalino iron oxide copper-gold deposit, in northern Brazil. Ref. [104] applies a new technique for airborne magnetic topography correction on data collected over the Río Blanco-Los Bronces and El Teniente porphyry copper districts (Andes of central Chile). Ref. [105] utilizes 3D litho-constrained inversion of potential field data to define a region with good potential for iron ore deposits in the Sierra Grande region of northern Patagonia, Argentina. Ref. [106] compares four techniques for removing spatial aliasing artifacts in aeromagnetic surveys completed over southeastern Minas Gerais, Brazil. Using the Poisson theorem relating magnetization and density, Ref. [107] interprets a pseudo potential field survey covering the Serra Sul of the Carajás Mineral Province, Brazil, area known for iron deposits. An analysis by [108] of the geophysical data collected at the Cristalino IOCG deposit in Carajás, Brazil, reveals that magnetic rocks are not fully mapped by the total magnetic gradient if the strike of the rocks direction and Earth's magnetic field direction coincide. Furthermore, brecciated massive sulfides are found to be weakly magnetic. Ref. [109] presents results of 2D modelling and 3D inversion of airborne magnetic data flown over the Serra das Éguas Complex, (in Brumado, Bahia, northeastern Brazil), a geological structure containing lenses of magnesite. Ref. [110] uses gravity and magnetic methods for the investigation of an iron ore deposit, namely East Deposit, of Sierra Grande Formation, Río Negro Province, Argentina.

3.7. Mixed Continents

Ref. [111] reviews the practice of gravity, magnetic, and radiometric methods to the study of igneous intrusions, notably carbonatitic–alkalic intrusions, peralkaline intrusions, and pegmatites, rich in rare-earth metals, with North American and Australian examples.

4. Electromagnetics

4.1. Africa

Ref. [112] compares four different AEM system data sets from the Sunnyside nickel deposit in southeast Botswana. Ref. [113] presents filtered and enhanced Airborne Electromagnetic (AEM) time domain data acquired over the Ilesha Shist Belt in southwestern

Nigeria. Ref. [114] presents a study, using ground geophysical methods, aimed at detecting sulfide mineralization possibly associated with radioactive bostonite rocks located in the Central Eastern Desert, Egypt.

4.2. Asia

Ref. [115] describes the results of high-resolution heliborne magnetic and time domain EM surveys conducted over a fracture-controlled uranium deposit located in the Sikar district, Rajasthan, India. Ref. [116] describes the results of a controlled source audio-frequency magnetotelluric (CSAMT) survey carried out to evaluate potential iron (Fe) and polymetallic (Pb-Zn-Cu) deposits in Longmen region, southern China. Ref. [117] attributes to polarizable conductive zones negative responses in heliborne EM data recorded over uranium deposits in the Cuddapah Basin, India. Ref. [118] evaluates a chromite exploration case history using CSAMT, located in southern Tibet, China. Ref. [119] compares broadside (perpendicular) and inline ground time domain EM methods over the Caosiyao porphyry molybdenum deposit in China. Ref. [120] presents 3D MT inversions in the central Luzong basin ore district of eastern China which is well known for porphyry iron deposits. Ref. [121] describes the results of an AMT survey conducted over a gold polymetallic deposit in the Qinling Metallogenic Belt of North China Craton. Ref. [122] 3D inverts AMT data over the Cimabanshuo porphyry copper deposit in Tibet. Ref. [123] analyzes the application of the AMT technique to a metallic deposit located in the northwestern Guizhou province of China.

4.3. Australia

Ref. [124] compares 3D inversions of AEM systems ZTEM and AirMT collected over the Nebo-Babel Ni-Cu-PGE deposit, in West Musgrave, Western Australia. Ref. [125] presents airborne EM results of surveys over known and new manganese deposits located in Western Australia. Ref. [126] presents results of airborne and borehole EM surveys at Hallandaire copper deposit in Western Australia. Ref. [127] integrates inverted AEM data with the initial model of a 2D MT inversion conducted over the Cariewerloo Basin, a region of unconformity-type uranium deposit, in South Australia. Ref. [128] presents and interprets sub-audio magnetics data over the Far South gold project located in Western Australia. Ref. [129] describes airborne and ground EM surveys which led to the discovery of the Musket and Camelwood nickel sulfide deposits located in Western Australia. Ref. [130] presents drill hole EM surveys that led to the discovery of the Eureka massive sulfide lens located in Victoria. Ref. [131] describes the discovery of the Artemis Cu-Au-Zn-Ag deposit, located in northwest Queensland, using historical geophysical and geological data complimented by new airborne and ground EM. Ref. [132] presents and discusses AEM results over banded iron formation in the Hamersley Province of Western Australia. Ref. [133] attributes second-order effects in concentric-loop AEM system responses at Lewis Ponds in New South Wales, Australia, to polarizable material mapped with a Cole–Cole model as ground dipole–dipole array data accurately imaged disseminated sulfides surrounding ore-grade massive sulfides. Ref. [134] compares AEM responses collected over the Nebo Babel Ni-Cu-PGE deposit located in Western Australia. Ref. [135] delineates cobalt targets using the sub-audio magnetics method deployed over cobalt high-grade mineralization of the Carlow Castle project of Western Australia. Ref. [136] compares three inversion methods applied to greenfield AEM data. CDI3D, Maxwell codes, and underdetermined 3D voxellated inversion were used over a volume surrounding the conductor at the test site in Forrestania, Western Australia. The first two methods provide satisfactory results in comparison with the third method, which computes using significant computer resources but without satisfactory results. Ref. [137] presents results of borehole EM surveys associated with the mineralization in the Bellevue Gold Mine in the Agnew-Wiluna greenstone belt. Ref. [138] presents observations and interpretations from an AEM survey conducted over Mississippi valley-type lead-zinc (MV Pb-Zn) sulfide occurrences of the Canning Basin. Ref. [139] presents AEM data collected over PGE-Ni-Cu sulfide zones of the Julimar

Region of Western Australia. Ref. [140] presents the application of airborne passive EM for targeting hidden mineralization near the Telfer gold–copper deposit. Ref. [141] describes a discovery of nickel sulfide mineralization in the Yilgram region found by recovering AEM data's weakly conductive features, originally masked by IP effects.

4.4. Europe

Ref. [142] describes seismic reflection and MT surveys carried out in northwestern Skellefte District, Sweden, with the objectives of explaining the geological relationship between the mineralizations in the Adak mining camp and the Kristineberg area. Ref. [143] describes ground electric and EM surveys conducted over the Pb-Zn deposit of Lontzen-Poppelsberg, Belgium. Ref. [144] describes audio 2D MT inversion results from data collected over the Outokumpu belt in eastern Finland, characterized by several polymetallic (Cu-Co-Zn-Ni-Ag-Au) sulfide ore deposits. Ref. [145] presents the results of ground-based long-offset transient-electromagnetic data in eastern Thuringia, Germany. Plate models derived for AEM data are compared by [146] with drilled sections of the massive sulfide volumes from the Touro copper deposit in Spain, which showed excellent correlation. Ref. [147] presents results from ground electrical and electromagnetic surveys carried out at Koillismaa Layered Intrusion Complex in northeastern Finland. Ref. [148] presents a Deep EM Sounding for Mineral Exploration (DESMEX) investigation in a graphite mining district in eastern Bavaria, Germany, where IP effects were observed in 3D inversions.

4.5. North America

Ref. [149] compares resistivity maps estimated from ground and various AEM surveys conducted over the Midwest uranium deposit located in Saskatchewan, Canada. Ref. [150] contrasts SPECTREM fixed-wing and ZTEM natural audio-frequency inversion results from surveys flown over Pebble deposit in Alaska, USA. Ref. [151] presents the results of helicopter natural EM field measurements over the low sulfidation epithermal gold–silver vein systems at Gold Springs located in southeastern Nevada, USA. Ref. [152] presents inversion and interpretation of airborne natural EM field data over the Silver Queen polymetallic vein system located in BC, Canada. Ref. [153] compares results and interpretations of passive and time-domain helicopter surveys flown over the 501 project (Cu-Zn volcanogenic massive sulfide) at McFault's Lake in northern Ontario, Canada. Ref. [154] presents airborne, ground, and borehole EM measurements over the Hood 10 volcanogenic hosted massive sulfide located in Nunavut, Canada. Ref. [155] analyzes the response of ground electromagnetic surveys conducted over the former Opemiska mine in Québec, Canada, where petrophysical studies, geological and geophysical data have been compiled. Ref. [156] presents a case study of a ZTEM survey conducted over sedimentary exhalative (SEDEX) lead-zinc deposits in Selwyn Basin, Yukon, Canada. Ref. [157] describes the results of the HELITEM system flown over the Lalor deposit in Manitoba, Canada. Ref. [158] investigates nine inversion strategies applied to co-located seismic and MT data sets from one of the Carlin gold deposit districts of north-central Nevada, USA. Ref. [159] analyzes AEM data for Lac Brûlé, Québec, Canada, over an anorthosite intrusion, and is able in some areas to distinguish, in the data, superparamagnetic (SPM) effects from IP effects. Ref. [160] compares 3D inversions of ZTEM and ground MT data over a Ag-Au-rich epithermal system in Canada, while Ref. [161] compares the same AEM system inversions over the Morrison porphyry Cu-Au-Mo deposit, both of which are located in British Columbia, Canada. Ref. [162] presents results of AEM time-domain and natural field data over the Dolly Varden Mine region of BC, which hosts silver and gold deposits. Ref. [163] presents airborne and ground, radiometric and EM, surveys collected over the Patterson Lake South uranium deposit located in Saskatchewan, Canada. Trial-and-error interpretation is attempted on EM data from the uranium rich Athabasca Basin, Saskatchewan [164,165]. Ref. [166] presents an interpretation of AEM methods using the moment Gaussian model developed by [167] from the Athabasca Basin, Canada. Ref. [168] presents a study of forward modelling and 3D EM inversion from the McArthur

River uranium deposit in the Athabasca Basin, Canada, integrating the finding about the quaternary geological cover impact on EM measurements.

4.6. South America

Ref. [169] presents 3D audiomagnetotelluric inversions over the Regis kimberlite pipe in Minas Gerais, Brazil. Ref. [170] evaluates a method of cooperative inversion on airborne TEM, controlled source magnetotellurics, and direct current resistivity data from the Antonio gold deposit in Peru. Ref. [171] presents Airborne Induced Polarization (AIP) inversions for data collected over the Lamego gold mine of the Quadrilátero Ferrífero area in Minas Gerais State, Brazil. Ref. [172] presents a review of historical ground and downhole EM data, identifying strong correlations with known nickel intersects from the Jaguar Nickel Project, a hydrothermal Iron-Oxide Nickel Copper (IONC) deposit located in the Carajas Mineral Province of Brazil.

4.7. Mixed Continents

Ref. [173] outlines an approach to extract IP information from AEM data using the Cole–Cole model on VTEM data collected from Mt Milligan, BC, Canada and Amakinskaya kimberlite pipe, Russia. Ref. [174] interprets in terms of Cole–Cole parameters AIP surveys conducted over various geologies and exploration targets, with examples of diamond, gold, and base metals deposits in northern Canada, Oman, and Australia. Ref. [175] compares 1D and 2D AEM inversion algorithms applied over areas of rough terrains in Iran and Norway. Ref. [176] presents case studies of application of AEM natural fields in exploring for various minerals across sites located in Canada and Kazakhstan.

5. Magnetics and Electromagnetics

5.1. Africa

Ref. [177] presents a robust airborne natural source processing approach, which is applied to data from the Kalahari-Copper-Belt in Namibia.

5.2. Asia

Ref. [178] describes geological mapping and detailed geophysical surveys for chromite exploration at Tangarparha, India. Ref. [179] presents airborne and ground geophysical measurements over the Elang Cu-Au porphyry deposit in Indonesia. Ref. [180] presents results of helicopter natural EM fields and magnetics flown over the Ad Duwayhi intrusion-related gold deposit in Saudi Arabia. Ref. [181] presents advanced processing and interpretation, for AEM, magnetic, and radiometric data from the Mohar cauldron, India, with uranium potential. Ref. [182] compares magnetic and electromagnetic (EM) results from time-domain (VTEM) and AFMAG (ZTEM) flown over the Nuqrah SEDimentary EXhalative (SEDEX) massive sulfide deposits located in the Kingdom of Saudi Arabia. Ref. [183] presents the results of audio MT and ground magnetic surveys which help image the alteration and mineralization system of the epithermal gold deposit, Tuoniuhe, in northeast China. Ref. [184] presents results of a comprehensive analysis of 2D CSAMT imaging over an important copper polymetallic ore field in Zhongxingtun area, Inner Mongolia, China. From 2D profile CSAMT inversions, Ref. [185] investigates the electrical resistivity structure of the Zhaishang gold deposit, in West Qinling, China. Ref. [186] combines magnetics, TEM, and magnetotellurics to locate an iron ore body from southern Liaoning province, China. Ref. [187] uses potential field and electromagnetic data interpretation to provide new information on the copper-zinc Xiaorequanzi deposit, in northwest China. Ref. [188] presents a study analyzing magnetic and VLF-EM data to delineate anomalous zones related to gold-associated sulfide mineralization in the North Singhbhum Mobile Belt of eastern India. Ref. [189] presents results of ground magnetic and VLF-EM surveys conducted over the Hatinitor Pahar and Kadwara areas of India, which are probable kimberlite zones.

5.3. Australia

Ref. [190] reviews the various geophysical methods that have been tested over the high grade Ag-Zn-Pb Cannington deposit located in Queensland. Ref. [191] reviews integrated interpretation of geophysical surveys with geological data applied in the interpretation and exploration of detrital iron deposits in Western Australia. Ref. [192] describes ground EM, magnetic, and gravity surveys obtained over the Gurubang volcanogenic massive sulfide (VMS) deposit located in New South Wales. Ref. [193] present results of ground and airborne surveys conducted over the Wafi-Golpu porphyry system.

5.4. Europe

Ref. [194] analyzes the geophysical characteristics of a strongly-conductive horizon in northern Sweden, associated with thin layers of pyrrhotite and graphite. Ref. [195] derives a 3D conductivity volume from 2D MT-inverted sections of data collected over the Kevitsa Ni-Cu-PGE deposit in Finland. Ref. [196] develops a new boundary detection technique that identifies and removes uniform layers from the 3D model, leaving edges. The method allows for the visualization of property edges, which are generated from the geophysical inversion of data from the Kevitsa Ni-Cu-PGE deposit in Lapland, Finland.

5.5. North America

Ref. [197] presents an application of lithoclassification using the self-organizing maps (SOM) technique, which uses EM, gravity and magnetic data from the Reid–Mahaffy test site in Ontario, Canada. Ref. [198] characterizes the Cu-Au-Mo Pebble porphyry deposit in Alaska, USA, using geophysical data, drillhole information, and physical properties.

Ref. [199] presents results and preliminary interpretation of AEM and magnetic surveys over the massive sulfide Lalor deposit in Manitoba, Canada, while [200,201] present inversion results of EM and airborne magnetic data collected over the same deposit. Ref. [202] presents an integration of inversion of aeromagnetic, gravity, and MT data with physical property measurements and geological information, thus facilitating a better understanding of the NICO Au-Co-Bi-Cu deposit, Northwest Territories, Canada. Ref. [203] combines MT, magnetics, and gravity to study the geophysical response of the Mountain Pass carbonatite, in California.

Ref. [204] presents results of joint inversions of two VTEM datasets and associated aeromagnetic datasets, from two surveys flown six years apart over a VMS gold prospect in northern Ontario, Canada. Results of inversions over Tli Kwi Cho kimberlites from Northwest Territories, Canada, are presented for potential field data [205], airborne AEM conductivity [206] and chargeability data [207]. Ref. [208] compiles historical apparent resistivity and induced polarization data combined with recent AEM surveys to infer structurally complex zones in the gold-rich, Canadian Malartic district, from Québec, Canada. Ref. [209] uses a supervised machine learning algorithm to classify rock types throughout the Kliyul porphyry prospect in British Columbia, Canada. Ref. [210] inverts magnetic and frequency-domain EM data over gold-rich, Canadian Malartic district, and interprets the results in terms of lithologies. Ref. [211] presents modelling and analysis of helicopter EM data resulting in physical property models and magnetic derivatives which characterize shallow parts of the Stillwater Complex, USA, rich in PGE, nickel, copper, and chromium. Ref. [212] describes data, interpretation, and targeting approach of a helicopter time domain EM, magnetic, and radiometric survey over epithermal Au-Ag deposits in north central BC, in Canada. Ref. [213] describes the results of airborne natural EM field and magnetic survey conducted over porphyry copper occurrences near Houston, also in BC.

5.6. South America

Ref. [214] compares ground resistivity/IP, airborne EM, and magnetic responses, supported by 3D inversions, over the Romero and Romero South Au-Cu-Zn-Ag epithermal deposit area of Dominican Republic. Ref. [215] describes the data and interpretation of an airborne EM, magnetic, and radiometric survey flown over the Cerro Quema high sulfidation epithermal gold deposits in Panama. Ref. [216] describes airborne EM and magnetic survey results from the Panama Cobre porphyry copper camp, which includes the Balboa deposit.

6. Discussion

The numerous references presented show that magnetic and EM techniques are still popular for mineral exploration and that this field represents an active area of research with many papers published in various scientific journals. In particular, drone magnetic surveys have gained popularity. We have compiled the different case studies (drone studies excluded) by target type and continents for the magnetic technique in Table 1, the EM technique in Table 2, and a combination of magnetic and EM techniques in Table 3.

Table 1. Compilation of case studies by application of magnetics (IOCG: Iron-Oxide Copper-Gold, MS: Massive Sulfides).

Deposit Type	Africa	Asia	Australia	Europe	North America	South America
Apatite		[50]			[93]	
Carbonatite			[111]		[96,111]	[101]
Copper		[51,55]				[102]
Gemstones	[41]			[77]		
Gold	[38,42,45,47]	[49,58]	[63,64,67,68,71,72]		[87,88,95]	[97]
IOCG		[54]	[61,65,70]			[100,103,108]
Iron ore		[52,53,56]	[62]	[75]		[99,105–107,110]
Kimberlite	[35,37,46]				[91,92]	
Lateritic Ni						[98]
Magnesite						[109]
Manganese	[43]		[66]			
MS			[74]			
Ni-Cu-PGE					[81,83–85]	
Phosphate	[44]					
Polymetallic	[36,40]	[48,59,60]	[73]	[76]	[78,79,94]	
Porphyry Cu		[57]	[69]		[82,88–90]	[104]
Rutile	[39]					
Uranium					[80,86,88]	

Table 2. Compilation of case studies by application of electromagnetics (IONC: Iron-Oxide Nickel-Copper, MS: Massive Sulfides, MVLZ: Mississippi Valley Lead-Zinc, SEDEX: SEDimentary EXhalative lead-zinc).

Deposit Type	Africa	Asia	Australia	Europe	North America	South America
Bostonite	[114]					
Chromite		[118]				
Cobalt			[135]			
Copper			[126]	[146]		
Gold	[113]	[121]	[137,140,174]		[151,158,174]	
Gold & zinc					[156,157]	[170,171]
Graphite				[148]		
IONC						[172]
Iron ore		[120]	[132]			
Kimberlite		[173]			[173]	[169]
Manganese			[125]			
Molybdenum		[119]				
MS			[128,130,133,136]	[142]	[153–155]	
MVLZ			[138]	[143]		
Ni-Cu-PGE	[112]		[124,129,134,139,141]	[147]		
Polymetallic		[116,123]	[131]	[144]	[152]	
Porphyry Cu		[122]			[150,161]	
Rutile					[159]	
SEDEX					[156]	
Silver					[160,162]	
Uranium		[115,117]	[127]	[145]	[149,163–165,167,168]	

Table 3. Compilation of case studies on application by magnetics and electromagnetics (IOCG: Iron-Oxide Copper-Gold, MS: Massive Sulfides, SEDEX: SEDimentary EXhalative lead-zinc).

Deposit Type	Africa	Asia	Australia	Europe	North America	South America
Carbonatite					[203]	
Chromite		[178]				
Copper	[177]					
Copper-zinc		[187]				
Detrital iron			[191]			
Epithermal gold					[212]	[214,215]
Gold		[180,183,185,188]			[208,210]	
Graphite				[194]		
IOCG					[202]	
Iron ore		[186]				
Kimberlite		[189]			[205–207]	
MS			[192]		[197,199–201,204]	
Ni-Cu-PGE				[195,196]	[211]	
Polymetallic		[184]				
Porphyry Cu		[179]	[193]		[198,209,213]	[216]
SEDEX		[182]				
Silver			[190]			
Uranium		[181]				

These results can be used to produce various histograms and only the most meaningful are presented. The simplest one presented in Figure 1 shows the number of research papers for each method while Figure 2 shows the number of papers by continent. The majority of case studies are compiled for the magnetic method and describes surveys conducted in North America, followed by Asia and Australia, while other continents have smaller but similar numbers of papers. The popularity of the magnetic technique may be a reflection of the fact that this technique is used both as a direct and an indirect method of exploration. Magnetic surveys can be conducted to directly detect magnetic material like iron ore, while during the search of other deposits they are used mainly to map the geology, using the rock magnetite content as a geological indicator or structural marker. In general, EM surveys directly detect the conductive mineralization, although in uranium exploration they detect conductors associated with the mineralization.

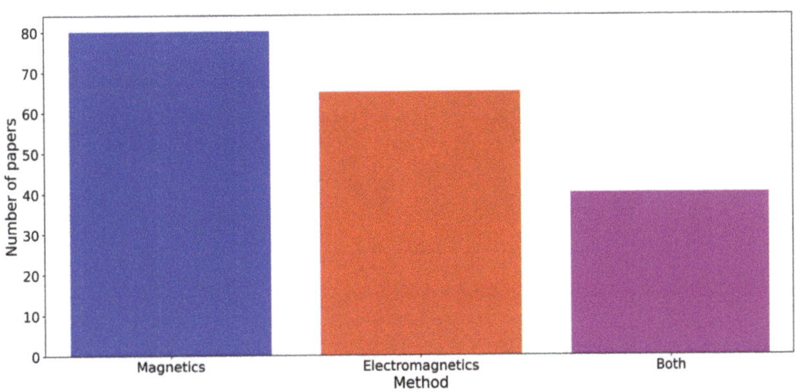

Figure 1. Number of case studies by method.

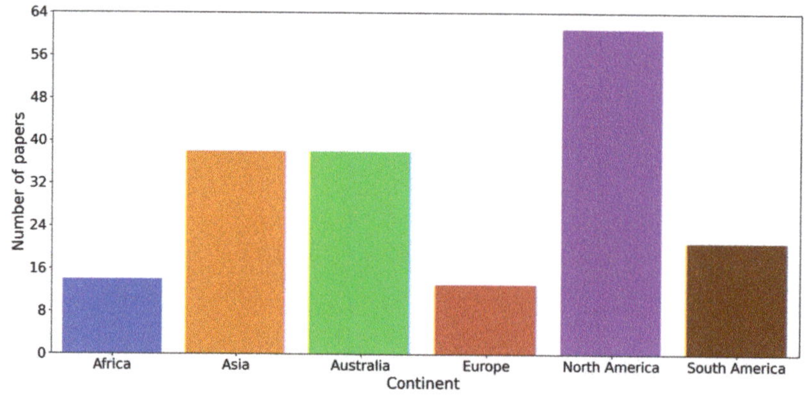

Figure 2. Number of case studies by continent.

Going into more detail, an histogram of the number of papers by method and continent (Figure 3) shows distinct patterns. Electromagnetic and magnetic paper numbers are the same in North America, while magnetic methods are more popular in other continents except Europe, where EM methods are favored. The combination of magnetic and EM techniques is popular in North America and Asia. Although the number of case studies documented in a particular continent is function of various factors such as the geography and metallogeny, the mining market, the exploration activity, the number of researchers in each continent, and the bias of the scientific journals that were used for this study, it is, therefore, a first-order indication of the preference for each method in these continents.

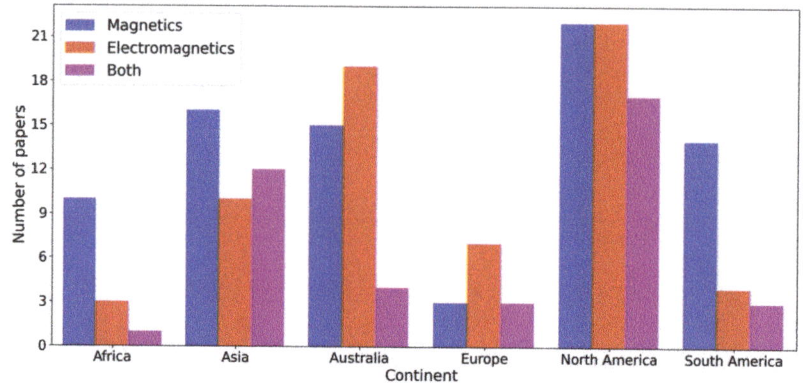

Figure 3. Number of case studies by continent and method.

The following figures illustrate the distribution of the various deposit types documented. In Figure 4, the total number for each target type is presented. The classification of the deposits was based on the information provided by the authors of the case studies and is sometimes more commercial than geological. However, characteristic features can be observed. Gold mineralizations are the most popular, followed by porphyry copper, polymetallic, massive sulfides, uranium, Ni-Cu-PGE, iron ore, kimberlite, and IOCG.

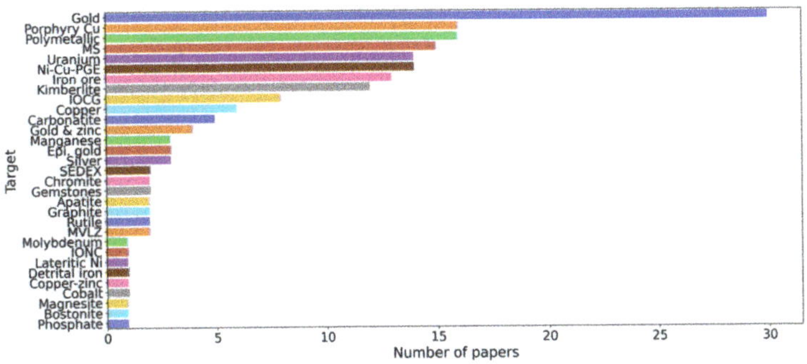

Figure 4. Number of case studies by types of mineralization (IOCG: Iron-Oxide Copper-Gold, IONC: Iron-Oxide Nickel-Copper, MS: Massive Sulfides, MVLZ: Mississippi Valley Lead-Zinc, SEDEX: SEDimentary EXhalative lead-zinc).

In Figure 5, these types of mineralization are classified by methods. Magnetic techniques are favored for gold, iron ore, and polymetallic, while electromagnetic methods are used for uranium, massive sulfides, and gold. Both methods characterize porphyry copper, massive sulfides, and gold.

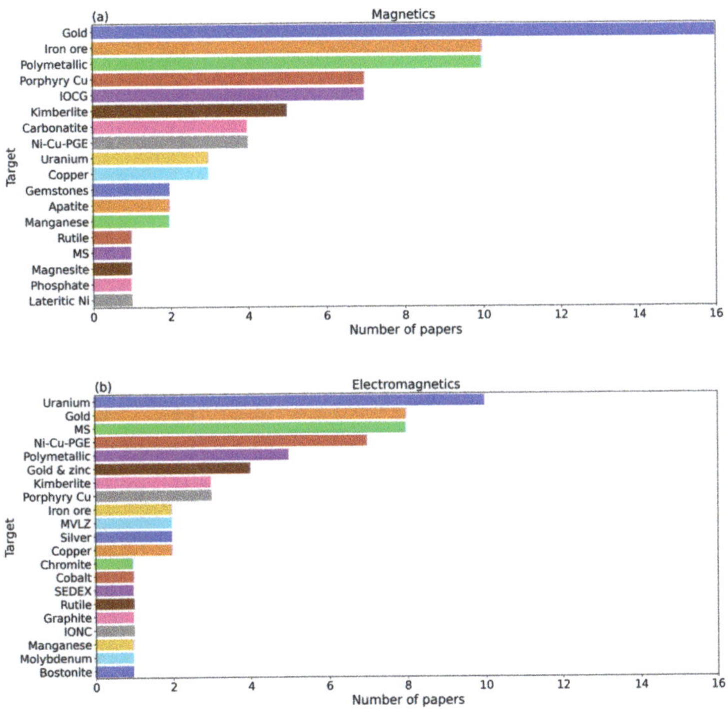

Figure 5. *Cont.*

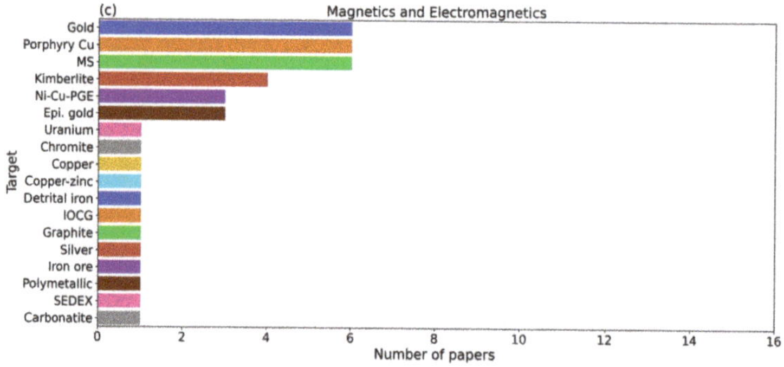

Figure 5. Types of mineralization by method: (**a**) magnetics, (**b**) electromagnetics, and (**c**) magnetics and electromagnetics (IOCG: Iron-Oxide Copper-Gold, IONC: Iron-Oxide Nickel-Copper, MS: Massive Sulfides, MVLZ: Mississippi Valley Lead-Zinc, SEDEX: SEDimentary EXhalative lead-zinc).

Figure 6 presents the same information by continents. Gold and polymetallic-style mineralization are studied in Africa, while gold, polymetallic, and kimberlite are studied in Asia. In Australia, gold and massive sulfides are studied geophysically, while in Europe, case studies for Ni-Cu-PGE, polymetallic, and graphite are more popular. In North America, studies have been conducted predominantly on uranium, porphyry copper, massive sulfides, gold, kimberlite, and others. In South America, iron ore, IOCG, and porphyry copper mineralizations have been studied.

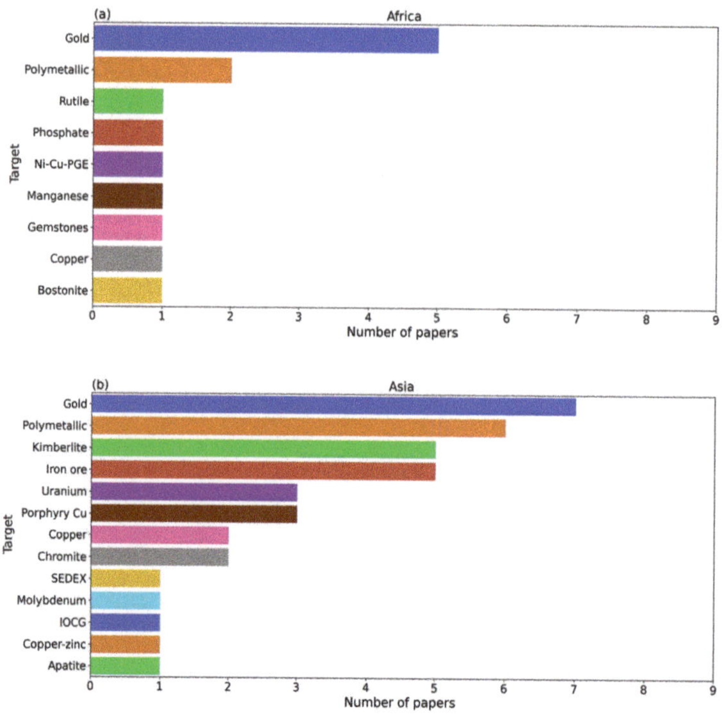

Figure 6. *Cont.*

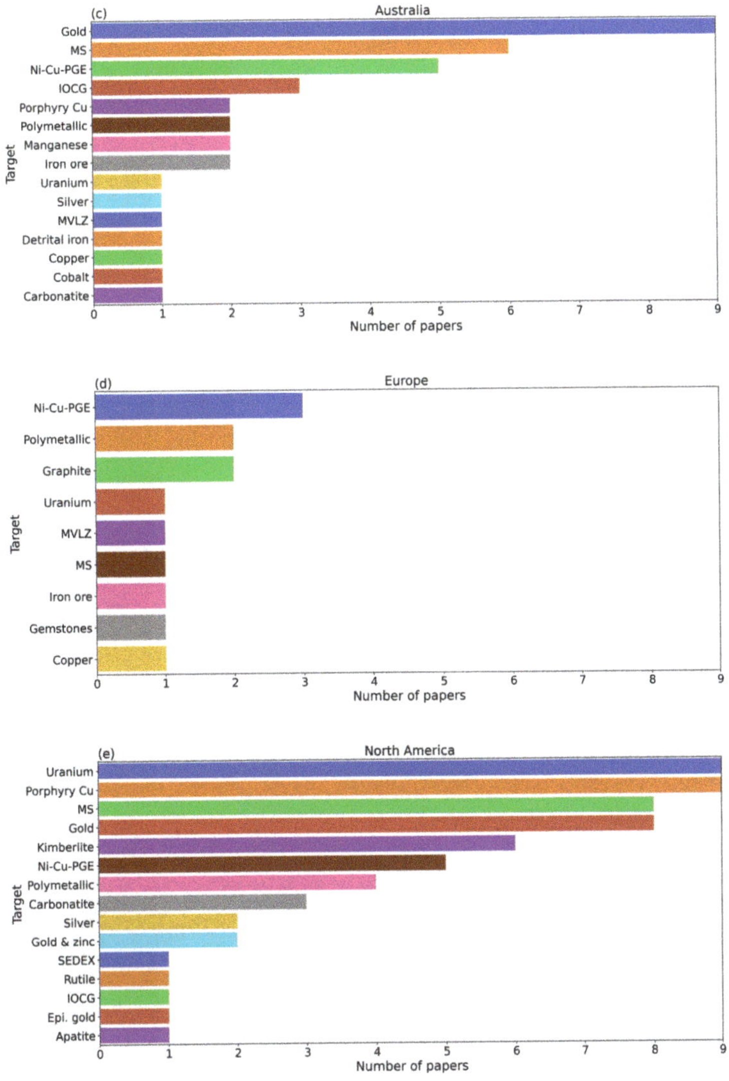

Figure 6. *Cont.*

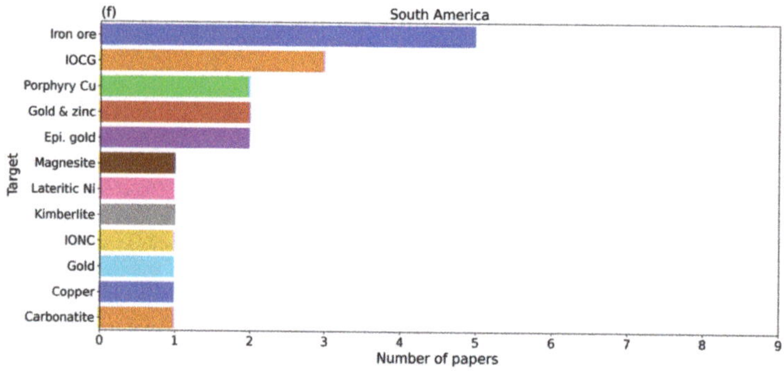

Figure 6. Types and numbers of mineralizations by continent: (**a**) Africa, (**b**) Asia, (**c**) Australia, (**d**) Europe, (**e**) North America, and (**f**) South America (IOCG: Iron-Oxide Copper-Gold, IONC: Iron-Oxide Nickel-Copper, MS: Massive Sulfides, MVLZ: Mississippi Valley Lead-Zinc, SEDEX: SEDimentary EXhalative lead-zinc).

7. Conclusions

The field of geophysical exploration for mineral deposits is very active with a constant flow of new developments, in particular in magnetic and EM techniques, as developers have been eager to adapt new technologies to new geophysical applications. Our review of the documented case studies over the last fifteen years is unfortunately an incomplete compilation of the geophysical activity around the world during this period, as not all geophysical activity is documented in the reviewed papers. First, case studies provide an incomplete description of the technological developments. Furthermore, in our review, applications of the magnetic and electromagnetic techniques in North America, Asia, and Australia predominate over applications in other continents. This predominance is attributed to different factors, ranging from academic and metallogeny, to mining market, and journals selected. Despite these limitations, this review provides detailed information about the usage of magnetic and EM techniques, encompassing a wide range of geological environments. The magnetic method is slightly more popular and this popularity is attributed to the fact that it can be both a direct and an indirect indicator of mineralization. Magnetic and EM techniques were utilized for many different target types, with a preference for gold, porphyry copper, polymetallic, massive sulfides, uranium, Ni-Cu-PGE, iron ore, kimberlite, and iron-oxide copper-gold. They are also popular for a range of small applications which are continent-specific. The variety of case studies presented in this review should help in selecting and designing geophysical surveys in a new area or domain as well in helping to better elucidate the challenges encountered in a specific geological environment, given the limitations of each technique.

Author Contributions: Conceptualization, M.A.V. and M.M.; methodology, M.A.V.; validation, M.A.V., M.M. and K.K.; investigation, M.A.V.; resources, M.A.V.; writing—original draft preparation, M.A.V.; writing—review and editing, M.A.V., M.M. and K.K.; visualization, M.A.V.; supervision, M.M.; project administration, M.M. All authors have read and agreed to the published version of the manuscript.

Funding: This research received no external funding.

Institutional Review Board Statement: Not applicable.

Informed Consent Statement: Not applicable.

Data Availability Statement: Not applicable.

Conflicts of Interest: Marc A. Vallée, Mouhamed Moussaoui and Khorram Khan are employees of Geo Data Solutions GDS Inc. The paper reflects the views of the scientists and not the company.

References

1. Nabighian, M.N.; Grauch, V.J.S.; Hansen, R.O.; LaFehr, T.R.; Li, Y.; Peirce, J.W.; Phillips, J.D.; Ruder, M.E. The historical development of the magnetic method in exploration. *Geophysics* **2005**, *70*, 33ND–61ND. [CrossRef]
2. Grant, F.S. Aeromagnetics, geology and ore environments, I. Magnetite in igneous, sedimentary and metamorphic rocks: An overview. *Geoexploration* **1985**, *23*, 303–333. [CrossRef]
3. Zhdanov, M.S. Electromagnetic geophysics: Notes from the past and the road ahead. *Geophysics* **2010**, *75*, 75A49–75A66. [CrossRef]
4. Palacky, G.J. Resistivity Characteristics of Geologic Targets. In *Electromagnetic Methods in Applied Geophysics: Volume I, Theory*; Nabighian, M.N., Ed.; Society of Exploration Geophysicists: Tulsa, OK, USA, 1987; pp. 53–129.
5. Smith, R. Electromagnetic Induction Methods in Mining Geophysics from 2008 to 2012. *Surv. Geophys.* **2014**, *35*, 123–156. [CrossRef]
6. S&P Global Market Intelligence. *The Big Picture 2023 Metals and Mining Industry Outlook*; S&P Global Market Intelligence: New York, NY, USA, 2022.
7. Reichl, C.; Schatz, M. *World Mining Data 2024*; Federal Ministry of Finance: Vienna, Austria, 2023.
8. Hammond, A.C.R. *The 20 Biggest Advances in Tech Over the Last 20 Years*; Foundation of Economic Education: Atlanta, GA, USA, 2020.
9. Palandrani, P. *A Decade of Change: How Tech Evolved in the 2010s an What's in Store for the 2020s*; Nasdaq: New York, NY, USA, 2022.
10. Vallée, M.A.; Smith, R.S.; Keating, P. Metalliferous mining geophysics—State of the art after a decade in the new millennium. *Geophysics* **2011**, *76*, W31–W50. [CrossRef]
11. Nabighian, M.N.; Asten, M.W. Metalliferous mining geophysics–State of the art in the last decade of the 20th century and the beginning of the new millennium. *Geophysics* **2002**, *67*, 964–978. [CrossRef]
12. Hodges, D.G.; Christensen, A.N. Airborne Geophysics. In Proceedings of the Exploration 17: Sixth Decennial International Conference on Mineral Exploration, Toronto, ON, Canada, 22–25 October 2017; pp. 5–16.
13. Fairhead, J.D.; Cooper, G.R.J.; Sander, S. Advances in Airborne Gravity and Magnetics. In Proceedings of the Exploration 17: Sixth Decennial International Conference on Mineral Exploration, Toronto, ON, Canada, 22–25 October 2017; pp. 113–127.
14. Macnae, J. Advances in Electromagnetic Data Processing: Noise, Signal, SPM and AIP. In Proceedings of the Exploration 17: Sixth Decennial International Conference on Mineral Exploration, Toronto, ON, Canada, 22–25 October 2017; pp. 191–208.
15. Macnae, J.; Hennessy, L. Magnetic Field Sensors for EM Geophysics. In Proceedings of the Exploration 17: Sixth Decennial International Conference on Mineral Exploration, Toronto, ON, Canada, 22–25 October 2017.
16. Farquharson, C.G.; Lelièvre, P.G. Modelling and Inversion for Mineral Exploration Geophysics: A Review of Recent Progress, the Current State-of-the-Art, and Future Directions. In Proceedings of the Exploration 17: Sixth Decennial International Conference on Mineral Exploration, Toronto, ON, Canada, 22–25 October 2017; pp. 51–74.
17. Jones, A.G. Magnetotellurics: Status Quo and Quo Vadimus. In Proceedings of the Exploration 17: Sixth Decennial International Conference on Mineral Exploration, Toronto, ON, Canada, 22–25 October 2017; pp. 139–158.
18. Chave, A.D.; Jones, A.G. (Eds.) *The Magnetotelluric Method: Theory and Practice*; Cambridge University Press: Cambridge, UK, 2018.
19. Duncan A. Advances in Ground and Borehole EM Survey Technology to 2017. In Proceedings of the Exploration 17: Sixth Decennial International Conference on Mineral Exploration, Toronto, ON, Canada, 22–25 October 2017; pp. 169–182.
20. Zhang, B.; Qu, K.; Yin, C.; Wang, Y.; Liu, Y.; Ren, X.; Su, Y. Review on 3D electromagnetic modeling and inversion for Mineral Exploration. *Geophysics* **2025**, *90*, WA47–WA59. [CrossRef]
21. Caron, R.M.; Samson, C.; Straznicky, P.; Ferguson, S.; Sander, L. Aeromagnetic surveying using a simulated unmanned aircraft system. *Geophys. Prospect.* **2014**, *62*, 352–363. [CrossRef]
22. Li, F.; Li, J.; Cui, Z.; Xu, Z. UAV-borne magnetic survey in Duobaoshan exploration area, Heilongjiang, China. In *International Workshop and Gravity, Electrical & Magnetic Methods and Their Applications, Chenghu, China, 19–22 April 2015*; Society of Exploration Geophysicists and and Chinese Geophysical Society: Houston, TX, USA, 2015; Volume 520.
23. Malehmir, A.; Dynesius, L.; Paulusson, K.; Paulusson, A.; Johansson, H.; Bastani, M.; Wedmark, M.; Marsden, P. The potential of rotary-wing UAV-based magnetic surveys for mineral exploration: A case study from central Sweden. *Lead. Edge* **2017**, *36*, 552–557. [CrossRef]
24. Cunningham, M.; Samson, C.; Wood, A.; Cook, I. Aeromagnetic Surveying with a Rotary-Wing Unmanned Aircraft System: A Case Study from a Zinc Deposit in Nash Creek, New Brunswick, Canada. *Pure Appl. Geophys.* **2018**, *175*, 3145–3158. [CrossRef]
25. Cunningham, M.; Samson, C.; Laliberté, J.; Goldie, M.; Wood, A.; Birkett, D. Comparison Between Ground, Helicopter, and Unmanned Aircraft System Magnetic Datasets: A Case Study from the Abitibi Greenstone Belt, Canada. *Pure Appl. Geophys.* **2022**, *179*, 1871–1886. [CrossRef]
26. Døssing, A.; Kolster, M.E.; Rasmussen, T.M.; Petersen, J.T.; da Silva, E.L.S. UAV-towed scalar magnetic gradiometry: A case study in relation to iron oxide copper-gold mineralization, Nautanen (Arctic Sweden). *Lead. Edge* **2023**, *42*, 103–111. [CrossRef]
27. Eremenco, D.; Parker, C.; Griffin, S.; Donaldson, M.; Cattach, M. Low-Frequency UAV SAM trials at Forrestania Electromagnetic Test Range in Western Australia. In Proceedings of the 1st ASEG DISCOVER Symposium, Hobart, Australia, 15–18 October 2024.
28. Walter, C.; Braun, A.; Fotopoulos, G. High-resolution unmanned aerial vehicle aeromagnetic surveys for mineral exploration targets. *Geophys. Prospect.* **2020**, *68*, 334–349. [CrossRef]
29. Cunningham, M.; Samson, C.; Laliberté, J.; Goldie, M.; Wood, A.; Birkett, D. Inversion of Magnetic Data Acquired with a Rotary-Wing Unmanned Aircraft System for Gold Exploration. *Pure Appl. Geophys.* **2021**, *178*, 501–516. [CrossRef]

30. Kim, B.; Jeong, S.; Bang, E.; Shin, S.; Cho, S. Investigation of Iron Ore Mineral Distribution Using Aero-Magnetic Exploration Techniques: Case Study at Pocheon, Korea. *Minerals* **2021**, *11*, 665. [CrossRef]
31. Zhou, W.; Zhang, C.; Tang, H.; Li, Q.; Tang, S. Iterative imaging method based on Tikhonov regularized downward continuation and its unmanned aerial vehicle aeromagnetic application: A case study from a Sijiaying iron deposit in eastern Hebei Province, China. *Geophysics* **2023**, *88*, B343–B354. [CrossRef]
32. Lee, M.; Madriz, Y.; Gloagen, R.; McEnroe, S. UAV magnetics over the Bjerkreim-Sokndal Intrusion, Rogaland, Norway: A first look. *Lead. Edge* **2023**, *42*, 90–97. [CrossRef]
33. Kaminski, V.; Davydenko, Y.; Bukhalov, S.; Goryachev, I. Combined ground-UAV TDEM survey over gold prospect in BaikalPatom Highlands (eastern Russia) with detection of AIP effect. In Proceedings of the 8th International Airborne Electromagnetics Workshop (AEM2023), Fitzroy Island, Australia, 4–7 September 2023.
34. Shahsavani, H.; Smith, R.S. Aeromagnetic gradiometry with UAV, a case study on small iron ore deposit. *Drone Syst. Appl.* **2024**, *12*, 1–9. [CrossRef]
35. Matende, K.; Mickus, K. Magnetic and gravity investigation of kimberlites in north-central Botswana. *Geophysics* **2021**, *86*, B67–B78. [CrossRef]
36. Ejiga, E.G.; Ismail, N.E.H.; Yusoff, I. Implementing Digital Edge Enhancers on Improved High-Resolution Aeromagnetic Signals for Structural-Depth Analysis around the Middle Benue Trough, Nigeria. *Minerals* **2021**, *11*, 1247. [CrossRef]
37. Vatankhah, S.; Renaut, R.A.; Mickus, K.; Liu, S.; Matende, K. A comparison of the joint and independent inversions for magnetic and gravity data over kimberlites in Botswana. *Geophys. Prospect.* **2022**, *70*, 1602–1616. [CrossRef]
38. El-Raouf, A.A.; Doğru, F.; Abdelrahman, K.; Fnais, M.S.; El Manharawy, A.; Amer, O. Using Airborne Geophysical and Geochemical Methods to Map Structures and Their Related Gold Mineralization. *Minerals* **2023**, *13*, 237. [CrossRef]
39. Diallo, S.; Ndam Njikam, M.M.; Yem, M.; Quentin Yene Atangana, J.; Ribodetti, A.; Raouf, A. Integration of magnetic and geological field data into geological mapping and rutile mineralization targets in the Minta locality (Haute-Sanaga, Cameroon). *Explor. Geophys.* **2023**, *54*, 670–684. [CrossRef]
40. Mbang, C.M.S.; Mouzong, M.P.; Cheng, L.Z.; Tafone, J.L.M.; Basseka, C.A.; Kamguia, J.; Etamè, J. Spatial Relations between Geological Structures and Precious and Base Metal Deposits from Magnetic Investigation of the Pangar-Djérem Zone, Cameroon. *Int. J. Geophys.* **2023**, *2023*, 4872249. [CrossRef]
41. Olawuyi, A.K. In-situ Location of Precious Stones using Magnetic Data and Pseudo-Gravity Transforms in Parts of the Nupe Basin Area of Nigeria. *SQU J. Sci.* **2024**, *29*, 86–94.
42. Sanusi, S.O.; Josiah, D.I.A.; Olaniyan, O.; Olayanju, G.M. Delineation of Potential Gold Mineralization Zones in the Kushaka Schist Belt, Northcentral Nigeria, Using Geochemical, Ground Magnetic, Induced Polarization, and Electrical Resistivity Methods. *Min. Metall. Explor.* **2024**, *41*, 2007–2029. [CrossRef]
43. Bahi, A.; Zerradi, Y.; Bziaz, M.; Tlidi, Y.; Soufi, A.; Bennouna, R.; Chakiri, S. Characterization of Mineral Meposits Using the Magnetic Method: The Case of the Aïn Beida Mine in Bouarfa, Morocco. *Civ. Environ. Eng.* **2024**, *20*, 571–579. [CrossRef]
44. Deep, M.A.; Ibrahim, A.S.; Saleh, A. Geophysical Assessment of Structurally Controlled Mineral Resources at Wadi El-Nakheel, Eastern Desert, Egypt. *Resources* **2024**, *13*, 83. [CrossRef]
45. Augie, A.I.; Salako, K.A.; Rafiu, A.A.; Jimoh, M.O. Integrated Geophysical Investigation for Potential Gold Mineralised Zone within Lower Part of Zuru Schist Belts, NW Nigeria. *Niger. J. Phys.* **2024**, *33*, 1–14. [CrossRef]
46. Vatankhah, S.; Lelièvre, P.G.; Matende, K.; Mickus, K. Magnetic surface geometry inversion of Kimberlites in Botswana. *Geophys. Prospect.* **2024**, *72*, 3524–3546. [CrossRef]
47. Harris, T.M.; Murphy, F.B. Geophysical Signatures of a +5.4Moz gold deposit, NE Bankan, Guinea, West Africa. In Proceedings of the ASEG DISCOVER Symposium (ASEG), Hobart, Australia, 15–18 October 2024.
48. Lü, Q.; Qi, G.; Yan, J. 3D geologic model of Shizishan ore field constrained by gravity and magnetic interactive modeling: A case history. *Geophysics* **2013**, *78*, B25–B35. [CrossRef]
49. Almasi, A.; Jafarirad, A.; Kheyrollahi, H.; Rahimi, M.; Afzal, P. Evaluation of structural and geological factors in orogenic gold type mineralisation in the Kervian area, north-west Iran, using airborne geophysical data. *Explor. Geophys.* **2014**, *45*, 261–270. [CrossRef]
50. Mandal, A.; Mohanty, W.K.; Sharma, S.P.; Biswas, A.; Sen, J.; Bhatt, A.K. Geophysical signatures of uranium mineralization and its subsurface validation at Beldih, Purulia District, West Bengal, India: A case study. *Geophys. Prospect.* **2015**, *63*, 713–726. [CrossRef]
51. Kheyrollahi, H.; Alinia, F.; Ghods, A. Regional magnetic lithologies and structures as controls on porphyry copper deposits: Evidence from Iran. *Explor. Geophys.* **2018**, *49*, 98–110. [CrossRef]
52. Sun, S.; Chen, C.; Liu, Y. Constrained 3D inversion of magnetic data with structural orientation and borehole lithology: A case study in the Macheng iron deposit, Hebei, China. *Geophysics* **2019**, *84*, B121–B133. [CrossRef]
53. Sunder Raju, P.V.; Satish Ku, K. Magnetic Survey for Iron-Oxide-Copper-Gold (IOCG) and Alkali Calcic Alteration Signatures in Gadarwara, M.P, India: Implications on Copper Metallogeny. *Minerals* **2020**, *10*, 671. [CrossRef]
54. Hasani, M.; Sharghi, Y.; Tercan, A.E. Separation of geomagnetic anomalies in the Bashmaq area in northwestern Iran by factorial kriging. *Explor. Geophys.* **2021**, *52*, 540–550. [CrossRef]
55. Riahi, S.; Bahroudi, A.; Abedi, M.; Aslani, S.; Elyasi, G.-R. Integration of airborne geophysics and satellite imagery data for exploration targeting in porphyry Cu systems: Chahargonbad district, Iran. *Geophys. Prospect.* **2021**, *69*, 1116–1137. [CrossRef]

56. Liu, Y.-X.; Li, W.-Y.; Liu, Z.-Y.; Zhao, J.-W.; Cao, A.-Q.; Gao, S.; Wang, L.-J.; Yang, C. Occurrence Characteristics of Magnetite and Aeromagnetic Prospecting Northeast of Hebei Province. *Minerals* **2022**, 12, 1158. [CrossRef]
57. Shabani, A.; Ziaii, M.; Monfared, M.S.; Shirazy, A.; Shirazi, A. Multi-Dimensional Data Fusion for Mineral Prospectivity Mapping (MPM) Using Fuzzy-AHP Decision-Making Method, Kodegan-Basiran Region, East Iran. *Minerals* **2022**, 12, 1629. [CrossRef]
58. Babu, S.B.; Patel, S.K.; Singh, R.B. An integrated geophysical approach for gold from Wayand-Nilambur granulite terrain, Kerala (India). *Geol. J.* **2024**, *59*, 1794–1811. [CrossRef]
59. Ji, N.; Qin, X.; Du, W.; Wang, Z.; Liu, Y.; Zhang, T. The Application of Aeromagnetic 3-D Geophysical Inversion in the Exploration of Lead–Zinc Polymetallic Deposits in Forested Areas. *IEEE Geosci. Remote Sens. Lett.* **2024**, *21*, 3433458. [CrossRef]
60. Zhang, H.; Li, H.; Hu, X. Fine interpretation of magnetic data for a concealed mineral deposit: A case study of the Fe-polymetallic deposit from Western China. *J. Appl. Geophys.* **2024**, *228*, 105468. [CrossRef]
61. Vella, L.; Cawood, M. Geophysical Characteristics of the Carrapateena Iron-Oxide Copper-Gold Deposit. *ASEG Ext. Abstr.* **2012**, *2012*, 1–4. [CrossRef]
62. Donohue, J.; Hill, Q.; Brewster, D. Geophysics at the Hawsons Iron Project, NSW—Eastern Australia's new magnetite resource. *ASEG Ext. Abstr.* **2012**, *2012*, 1–6. [CrossRef]
63. Clark, D.A. New methods for interpretation of magnetic vector and gradient tensor data II: Application to the Mount Leyshon anomaly, Queensland, Australia. *Explor. Geophys.* **2013**, *44*, 114–127. [CrossRef]
64. Clark, D.A. New methods for interpretation of magnetic vector and gradient tensor data I: Eigenvector analysis and the normalised source strength. *Explor. Geophys.* **2012**, *43*, 267–282. [CrossRef]
65. Austin, J.; Schmidt, P.; Lilly, R. Anisotropy of Magnetic Susceptibility (AMS) and Paleomagnetism applied to the differentiation of structural and metallogenic controls on Iron Oxide Copper-Gold (IOCG) mineralization: A case study from Monakoff, NW Queensland. *ASEG Ext. Abstr.* **2013**, *2013*, 1–5. [CrossRef]
66. Mosusu, N.T.; Verave, R.T.; Irarue, P.Y. Magnetic responses from an iron-rich gossan in a volcanogenic terrain and a limestone-hosted strata-bound manganese deposit, Central Province, Papua New Guinea. *ASEG Ext. Abstr.* **2013**, *2013*, 1–4.
67. Monoury, S.; Jupp, B.; Foley, A. Mine Scale Constrained Geophysical Inversion; A Case Study at the Darlot-Centenary Gold Mine. *ASEG Ext. Abstr.* **2015**, *2015*, 1–4. [CrossRef]
68. Banaszczyk, S.; Dentith, M.; Wallace, Y. Constrained Magnetic Modelling of the Wallaby Gold Deposit, Western Australia. *ASEG Ext. Abstr.* **2015**, *2015*, 1–4. [CrossRef]
69. Coote, A.; Erceg, M. Application of Petrology & Geology to the Interpretation of Geophysical Data in Defining Economic Porphyry-Related Cu-Au Mineralisation Along the Ekuti Range, Morobe Province, PNG. *ASEG Ext. Abstr.* **2016**, *2016*, 1–8.
70. Fabris, A.; Katona, L.; Reed, G.; Keeping, T.; Gordon, G.; Gouthas, G.; Swain, G. Mapping the Punt Hill IOCG system using geophysical, geochemical and spectral methods—An integrated approach. *ASEG Ext. Abstr.* **2016**, *2016*, 1–9. [CrossRef]
71. Kumwenda, J.; Lackie, M. Geophysical interpretation of the geology of the Stanthorpe region using aeromagnetic, gravity and radiometric data. *Explor. Geophys.* **2019**, *50*, 653–666. [CrossRef]
72. Pears, G.; Reid, J.; McGaughey, J.; Turner, W.; Hird, G. 3D interpretation of geological, 3D seismic and conventional geophysical data from the Darlot Gold Mine. *ASEG Ext. Abstr.* **2019**, *2019*, 1–5. [CrossRef]
73. Eshaghi, E.; Reading, A.M.; Roach, M.; Duffett, M.; Bombardieri, D.; Cracknell, M.J.; Everard, J.L.; Cumming, G.; Kuhn, S. Inverse modeling constrained by potential field data, petrophysics, and improved geologic mapping: A case study from prospective northwest Tasmania. *Geophysics* **2020**, *85*, K13–K26. [CrossRef]
74. Joly, A.; Reid, J.; Pears, G.; Paiement, J.-P.; Potter, D.; Healy, M.; Dunstone, K.; Hamill, J.; Nish, A. Integrated 3D modelling and associated machine learning targeting: The Jaguar Greenstone Belt example. In Proceedings of the Australasian Exploration Geoscience Conference (AEGC), Brisbane, Australia, 15–17 March 2023.
75. Aydin, I.; Uyanik, O.; Oksum, E.; Aydogan, M.S. Geophysical applications for Fe-rich emery exploration in the Elmacik area on the Menderes Massif (Turkey). *Explor. Geophys.* **2011**, *42*, 159–166. [CrossRef]
76. Kamm, J.; Lundin, I.A.; Bastani, M.; Sadeghi, M.; Pedersen, L.B. Joint inversion of gravity, magnetic, and petrophysical data—A case study from a gabbro intrusion in Boden, Sweden. *Geophysics* **2015**, *80*, B131–B152. [CrossRef]
77. Tavakoli, S.; Sarlus, Z.; Kronsell, I.; Bauer, T.E. 2.5D geophysical model of the Gällivare mining area: An integrated study to model the top 4 km of the subsurface and guide for future exploration activities. *Geophys. Prospect.* **2021**, *69*, 821–841. [CrossRef]
78. Spicer, B.; Morris, B.; Ugalde, H. Structure of the Rambler Rhyolite, Baie Verte Peninsula, Newfoundland: Inversions using UBC-GIF Grav3D and Mag3D. *J. Appl. Geophys.* **2011**, *75*, 9–18. [CrossRef]
79. Lee, M.; Morris, W.; Harris, J.; Leblanc, G. A network extraction tool for mineral exploration: A case study from the Wopmay Orogen, Northwest Territories, Canada. *Explor. Geophys.* **2012**, *43*, 116–124. [CrossRef]
80. Tschirhart, V.; Morris, W.A.; Jefferson, C.W.; Keating, P.; White, J.C.; Calhoun, L. 3D geophysical inversions of the north-east Amer Belt and their relationship to the geologic structure. *Geophys. Prospect.* **2013**, *61*, 547–560. [CrossRef]
81. Beiki, M.; Keating, P.; Clark, D.A. Interpretation of magnetic and gravity gradient tensor data using normalized source strength—A case study from McFaulds Lake, Northern Ontario, Canada. *Geophys. Prospect.* **2014**, *62*, 1180–1192. [CrossRef]
82. Anderson, E.D.; Zhou, W.; Li, Y.; Hitzman, M.W.; Monecke, T.; Lang, J.R.; Kelley, K.D. Three-dimensional distribution of igneous rocks near the Pebble porphyry Cu-Au-Mo deposit in southwestern Alaska: Constraints from regional-scale aeromagnetic data. *Geophysics* **2014**, *79*, B63–B79. [CrossRef]

83. Olaniyan, O.; Smith, R.S.; Lafrance, B. A constrained potential field data interpretation of the deep geometry of the Sudbury structure. *Can. J. Earth Sci.* **2014**, *51*, 715–729. [CrossRef]
84. Olaniyan, O.; Smith, R.S.; Lafrance, B. Regional 3D geophysical investigation of the Sudbury Structure. *Interpretation* **2015**, *3*, SL63–SL81. [CrossRef]
85. Mahmoodi, O.; Smith, R.S.; Spicer, B. Using constrained inversion of gravity and magnetic field to produce a 3D litho-prediction model. *Geophys. Prospect.* **2017**, *65*, 1662–1679. [CrossRef]
86. Tschirhart, V.; Pehrsson, S.J. New insights from geophysical data on the regional structure and geometry of the southwest Thelon Basin and its basement, Northwest Territories, Canada. *Geophysics* **2016**, *81*, B167–B178. [CrossRef]
87. Abbassi, B.; Cheng, L.Z.; Richards, J.P.; Hübert, J.; Legault, J.M.; Rebagliati, M.; Witherly, K. Geophysical properties of an epithermal Au-Ag deposit in British Columbia, Canada. *Interpretation* **2018**, *6*, T907–T918. [CrossRef]
88. Vallée, M.A.; Morris, W.A.; Perrouty, S.; Lee, R.G.; Wasyliuk, K.; King, J.J.; Ansdell, K.; Mir, R.; Shamsipour, P.; Farquharson, C.G.; et al. Geophysical inversion contributions to mineral exploration: Lessons from the Footprints project. *Can. J. Earth Sci.* **2019**, *56*, 525–543. [CrossRef]
89. Vallée, M.A.; Farquharson, C.G.; Morris, W.A.; King, J.; Byrne, K.; Lesage, G.; Enkin, R.J. Comparison of geophysical inversion programs run on aeromagnetic data collected over the Highland Valley Copper district, British Columbia, Canada. *Explor. Geophys.* **2019**, *50*, 310–323. [CrossRef]
90. Vallée, M.A.; Byrne, K.; King, J.J.; Lee, R.G.; Lesage, G.; Farquharson, C.G.; Chouteau, M.; Enkin, R.J. Imaging porphyry copper alteration using aeromagnetic data at Highland Valley Copper, British Columbia, Canada. *Explor. Geophys.* **2020**, *51*, 388–400. [CrossRef]
91. Darijani, M.; Farquharson, C.G.; Lelièvre, P.G. Joint and constrained inversion of magnetic and gravity data: A case history from the McArthur River area, Canada. *Geophysics* **2021**, *86*, B79–B95. [CrossRef]
92. Astic, T.; Fournier, D.; Oldenburg, D.W. Joint inversion of potential-fields data over the DO-27 kimberlite pipe using a Gaussian mixture model prior. *Interpretation* **2020**, *8*, SS47–SS62. [CrossRef]
93. Shah, A.K.; Taylor, R.D.; Walsh, G.J.; Phillips, J.D. Integrated geophysical imaging of rare earth element-bearing iron oxide-apatite deposits in the Eastern Adirondack Highlands, New York. *Geophysics* **2021**, *86*, B37–B54. [CrossRef]
94. Carter-McAuslan, A.; Farquharson, C. Predictive geologic mapping from geophysical data using self-organizing maps: A case study from Baie Verte, Newfoundland, Canada. *Geophysics* **2021**, *86*, B249–B264. [CrossRef]
95. Rimando, J.; Peace, A.L.; Geng, M.; Verbaas, J.; Slade, H. Structural Setting of the Sixtymile Gold District, Yukon, Canada: Insights into Regional Deformation and Mineralization from Field Mapping and 3D Magnetic Inversion. *Minerals* **2022**, *12*, 291. [CrossRef]
96. Wei, X.; Li, K.; Sun, J. Mapping critical mineral resources using airborne geophysics, 3D joint inversion and geology differentiation: A case study of a buried niobium deposit in the Elk Creek carbonatite, Nebraska, USA. *Geophys. Prospect.* **2023**, *71*, 1247–1266. [CrossRef]
97. de Carvalho Carneiro, C.; Fraser, S.J.; Crósta, A.P.; Silva, A.M.; de Mesquita Barros, C.E. Semiautomated geologic mapping using self-organizing maps and airborne geophysics in the Brazilian Amazon. *Geophysics* **2012**, *77*, K17–K24. [CrossRef]
98. Louro, V.H.A.; Mantovani, M.S.M.; Ribeiro, V.B. Magnetic field analysis of Morro do Leme nickel deposit. *Geophysics* **2014**, *79*, K1–K9. [CrossRef]
99. Martinez, C.; Li, Y. Lithologic characterization using airborne gravity gradient and aeromagnetic data for mineral exploration: A case study in the Quadrilátero Ferrífero, Brazil. *Interpretation* **2015**, *3*, SL1–SL13. [CrossRef]
100. Leão-Santos, M.; Li, Y.; Moraes, R. Application of 3D magnetic amplitude inversion to iron oxide-copper-gold deposits at low magnetic latitudes: A case study from Carajás Mineral Province, Brazil. *Geophysics* **2015**, *80*, B13–B22. [CrossRef]
101. Mantovani, M.S.M.; Louro, V.H.A.; Ribeiro, V.B.; Requejo, H.S.; Santos, R.P.Z.d. Geophysical analysis of Catalão I alkaline–carbonatite complex in Goiás, Brazil. *Geophys. Prospect.* **2016**, *64*, 216–227. [CrossRef]
102. Louro, V.H.A.; Mantovani, M.S.M.; Biondo Ribeiro, V.B. Integrated geologic and geophysical interpretation of the Buraco da Velha copper deposit (Rondônia, Brazil): A basis for exploring in related environments. *Geophysics* **2017**, *82*, B121–B133. [CrossRef]
103. Melo, A.T.; Sun, J.; Yaoguo, L. Geophysical inversions applied to 3D geology characterization of an iron oxide copper-gold deposit in Brazil. *Geophysics* **2017**, *82*, K1–K13. [CrossRef]
104. Yañez, G.; Ugalde, H.; Vargas, J.A. Topographic correction of magnetic data on rugged topography with application to Río Blanco-Los Bronces and El Teniente porphyry copper districts, Southern Andes, Chile. *Explor. Geophys.* **2018**, *49*, 595–607. [CrossRef]
105. Christiansen, R.; Kostadinoff, J.; Bouhier, J.; Martinez, P. Exploration of Iron ore deposits in Patagonia. Insights from gravity, magnetic and SP modelling. *Geophys. Prospect.* **2018**, *66*, 1751–1763. [CrossRef]
106. Fonte-Boa, T.M.R.; Melo, A.T.; Novo, T.A. Interpolation artifacts as a result of spatial aliasing: A case study of the airborne magnetic data set of southeastern Minas Gerais, Brazil. *Geophysics* **2020**, *85*, B193–B205. [CrossRef]
107. de Matos, C.A.; Mendonça, C.A. Poisson magnetization-to-density-ratio and magnetization inclination properties of banded iron formations of the Carajás mineral province from processing airborne gravity and magnetic data. *Geophysics* **2020**, *85*, K1–K11. [CrossRef]
108. Melo, A.T.; Li, Y.; Hitzman, M. Is there hidden potential in Carajás? Insights through the geophysical signature of Cristalino deposit. *Ore Geol. Rev.* **2020**, *126*, 103735. [CrossRef]

109. Sena, F.O.; Vasconcelos, M.A.R.; Pereira Cruz, S.C.; de Mendonça Freire, M.; Rangel, E.G.; Miranda, M.R. Magnetic modelling and physical properties of magnesite deposits and regional sequences of Serra das Éguas, Brumado, Brazil. *Geophys. Prospect.* **2023**, *71*, 336–349. [CrossRef]
110. Bahía, M.E.; Strazzere, L.; Benedini, L.; Gregori, D.A.; Kostadinoff, J. Gravity and magnetic exploration applied to iron ore deposits in the Sierra Grande area, Río Negro Province, Argentina. *Geophys. Prospect.* **2024**, *72*, 1175–1188. [CrossRef]
111. Thomas, M.D.; Ford, K.L.; Keating, P. Review paper: Exploration geophysics for intrusion-hosted rare metals. *Geophys. Prospect.* **2016**, *64*, 1275–1304. [CrossRef]
112. Viezzoli, A.; Selfe, G. Comparing responses from different AEM systems and derived models at the Sunnyside nickel project, Botswana. *ASEG Ext. Abstr.* **2018**, *2018*, 1–8. [CrossRef]
113. Osinowo, O.O.; Fashola, O.E.; Ayolabi, E.A.; Olayinka, A.I. Structural mapping and gold mineralisation potential evaluation from airborne time – domain electromagnetic (TDEM) data of Ilesha Schist Belt, southwestern Nigeria. *Explor. Geophys.* **2022**, *53*, 237–254. [CrossRef]
114. Abdelsalam, H.F.; Yousef, M.H.M.; Shaheen, M.A.; Badr, H.M.; El-Hawary, A.M.; Al-Alfy, I.M. Integration of geoelectric and electromagnetic methods to delineate subsurface mineralization in bostonite rocks, South El-Atshan area, Central Eastern Desert, Egypt. *Contrib. Geophys. Geod.* **2024**, *54*, 191–212. [CrossRef]
115. Chaturvedi, A.K.; Lotter, C.; Tripathi, S.; Maurya, A.K.; Patra, I.; Parihar, P.S. Integrated application of heliborne and ground electromagnetic surveys for mapping EM conductor for uranium exploration and its subsurface validation, North Delhi Fold Belt, Rajasthan, India: A case study. *Geophysics* **2013**, *78*, B13–B24. [CrossRef]
116. Hu, X.; Peng, R.; Wu, G.; Wang, W.; Huo, G.; Han, B. Mineral Exploration using CSAMT data: Application to Longmen region metallogenic belt, Guangdong Province, China. *Geophysics* **2013**, *78*, B111–B119. [CrossRef]
117. Babu, V.R.; Patra, I.; Tripathi, S.; Muthyala, S.; Chaturvedi, A.K. Inductive induced polarization effect in heliborne time-domain electromagnetic data for uranium exploration, northern part of Cuddapah Basin, India. *Geophysics* **2017**, *82*, B109–B120. [CrossRef]
118. He, L.; Chen, L.; Dorji; He, Z.; Wang, X.; Xiao, B.; Xu, L.; Zhao, X.; Xi, X.; Yao, H.; et al. Mapping chromite deposits with audio magnetotellurics in the Luobusa ophiolite of southern Tibet. *Geophysics* **2018**, *83*, B47–B57. [CrossRef]
119. Li, H.; Di, Q.-Y.; Xue, G.-Q. A comparative study of inline and broadside time-domain controlled-source electromagnetic methods for mapping resistive targets on land. *Geophysics* **2019**, *84*, B235–B246. [CrossRef]
120. Yan, J.; Zhang, K.; Jinhua, Z. Local uplift and copper metallogenic prospect in the Luzong ore district, Eastern China: Evidence from 3D magnetotelluric imaging. *Explor. Geophys.* **2019**, *50*, 22–30. [CrossRef]
121. Li, F.; Zeng, Q.; Zhu, R.; Chu, S.; Xie, W.; Zhang, B.; Zhang, X. Application of the AMT Method to Gold Deposits: A Case Study in the Qinling Metallogenic Belt of North China Craton. *Minerals* **2021**, *11*, 1200. [CrossRef]
122. Qi, P.; Yin, Y.; Jin, S.; Wei, W.; Xu, L.; Dong, H.; Huang, J. Three-Dimensional Audio-Magnetotelluric Imaging including Surface Topography of the Cimabanshuo Porphyry Copper Deposit, Tibet. *Minerals* **2021**, *11*, 1424. [CrossRef]
123. Pitiya, R.; Lu, M.; Chen, R.; Nong, G.; Chen, S.; Yao, H.; Shen, R.; Jiang, E. Audio Magnetotellurics Study of the Geoelectric Structure across the Zhugongtang Giant Lead–Zinc Deposit, NW Guizhou Province, China. *Minerals* **2022**, *12*, 1552. [CrossRef]
124. Legault, J.; Wilson, G.A.; Gribenko, A.V.; Zhdanov, M.S.; Zhao, S.; Kisk, K. An overview of the ZTEM and AirMt airborne electromagnetic systems: A case study from the Nebo–Babel Ni–Cu–PGE deposit, West Musgrave, Western Australia. *Preview* **2012**, *158*, 26–32. [CrossRef]
125. Kita, J.; Noetzli, A.; Kusumaputri, N.; Lowe, M. Airborne electromagnetics in Pilbara manganese exploration—A case study. *ASEG Ext. Abstr.* **2012**, *2012*, 1–2. [CrossRef]
126. Reid, J.; Price, D.; Summerhayes, E. Geophysical case history of the Hollandaire Copper Deposit, Western Australia. *ASEG Ext. Abstr.* **2013**, *2013*, 1–4. [CrossRef]
127. Crowe, M.; Heinson, G.; Dhu, T. Magnetotellurics and Airborne Electromagnetics—A combined method for assessing basin structure and exploring for unconformity-related uranium. *ASEG Ext. Abstr.* **2013**, *2013*, 1–5. [CrossRef]
128. Wallace, Y.; Card, D.; Gates, K.; Cattach, M. Identification of massive sulphide targets using the Galvanic Source EM (GSEM) signal from a Sub-Audio Magnetic (SAM) survey at the Far South Project, Western Australia. *ASEG Ext. Abstr.* **2015**, *2015*, 1–4. [CrossRef]
129. Huizi, A.; Mulholland, I.; Belbin, W. The Camelwood and Musket Nickel deposits - Discovery of a new Nickel Sulphide camp in the North-eastern Goldfields of Western Australia. *ASEG Ext. Abstr.* **2015**, *2015*, 1–4. [CrossRef]
130. Paggi, J.; Macklin, D. Discovery of the Eureka volcanogenic massive sulphide lens using down-hole electromagnetics. *ASEG Ext. Abstr.* **2015**, *2015*, 1–4. [CrossRef]
131. Thompson, A.D.; Flint, R.B. The discovery of the Artemis polymetallic deposit. *ASEG Ext. Abstr.* **2016**, *2016*, 1–6. [CrossRef]
132. Neroni, R.; Murray, R.; Kepert, D. Application of the airborne electromagnetic method for Banded Iron-Formation mapping in the Hamersley Province, Western Australia. *ASEG Ext. Abstr.* **2016**, *2016*, 1–8. [CrossRef]
133. Hine, K.; Macnae, J. Comparing induced polarization responses from airborne inductive and galvanic ground systems: Lewis Ponds, New South Wales. *Geophysics* **2016**, *81*, B179–B188. [CrossRef]
134. Witherly, K.; Sattel, D. An assessment of Geotem, Falcon® and ZTEM surveys over the Nebo Babel deposit, Western Australia. *ASEG Ext. Abstr.* **2018**, *2018*, 1–7. [CrossRef]
135. Eremenco, D.; Mortimer, R.; Mead, E. Delineating cobalt targets from a galvanic and inductive source Sub-Audio Magnetics (SAM) at the Carlow Castle project, Western Australia. *ASEG Ext. Abstr.* **2019**, *2019*, 1–4. [CrossRef]

136. Ren, X.; Macnae, J.; Hennessy, L. Three conductivity modelling algorithms and three 3D inversions of the Forrestania test site AEM anomaly. *Explor. Geophys.* **2020**, *51*, 14–24. [CrossRef]
137. Tomlinson, A.; Maude, G. Using modern downhole EM to discover high-grade, narrow vein lodes in a historical gold field, Bellevue Gold Project, Western Australia. *ASEG Ext. Abstr.* **2021**, *2021*, 1–4.
138. Zhan, Y.; Duuring, P.; Dent, L.; Brett, J. Electromagnetic response of MVT sulfide deposits on the Lennard Shelf, Canning Basin. In Proceedings of the Australasian Exploration Geoscience Conference (AEGC), Brisbane, Australia, 15–17 March 2023.
139. Burrows, D.; Smiarowski, A. Improving Nickel Exploration using a Purposefully Designed Airborne Electromagnetic System, with an Example from Julimar, Western Australia. *ASEG Ext. Abstr.* **2023**, *2023*, 1–4.
140. Prikhodko, A. Interpretation of electromagnetic multifrequency natural field data from Red Setter prospect (Paterson area, Western Australia). In Proceedings of the ASEG DISCOVER Symposium (ASEG), Hobart, Australia, 15–18 October 2024.
141. Neroni, R.; Blundell, K.; Viezzoli, A. Interpreting airborne electromagnetic data unburdened from induced polarisation effects: An unconventional mineral discovery case study from the eastern Yilgarn region of Western Australia. In Proceedings of the ASEG DISCOVER Symposium (ASEG), Hobart, Australia, 15–18 October 2024.
142. de los Ángeles García Juanatey, M.; Tryggvason, A.; Juhlin, C.; Bergström, U.; Hübert, J.; Pedersen, L.B. MT and reflection seismics in northwestern Skellefte Ore District, Sweden. *Geophysics* **2013**, *78*, B65–B76. [CrossRef]
143. Evrard, M.; Dumont, G.; Hermans, T.; Chouteau, M.; Francis, O.; Pirard, E.; Nguyen, F. Geophysical Investigation of the Pb–Zn Deposit of Lontzen–Poppelsberg, Belgium. *Minerals* **2018**, *8*, 233. [CrossRef]
144. Lahti, I.; Kontinen, A.; Nykänen, V. AMT survey in the Outokumpu ore belt, eastern Finland. *Explor. Geophys.* **2019**, *50*, 351–363. [CrossRef]
145. Mörbe, W.; Yogeshwar, P.; Tezkan, B.; Hanstein, T. Deep exploration using long-offset transient electromagnetics: Interpretation of field data in time and frequency domain. *Geophys. Prospect.* **2020**, *68*, 1980–1998. [CrossRef]
146. Núñez, P.; Watts, T.; Martin-Izard, A.; Arias, D.; Rubio, Á.; Cortés, F.; Díaz-Riopa, F. Airborne Electromagnetic Survey over the Touro Copper VMS World Class Deposit (NW Spain): Geological and Geophysical Correlation. *Minerals* **2023**, *13*, 17. [CrossRef]
147. Darnet, M.; Kim, B.; Védrine, S.; Deparis, J.; Bretaudeau, F.; Gance, J.; Vermeersch, F.; Truffert, C.; Autio, U.; Kamm, J.; et al. Ground electrical and electromagnetic methods for deep mineral exploration—Results from the SEEMS DEEP project. In Proceedings of the NSG 2024 5th Conference on Geophysics for Mineral Exploration and Mining, Helsinki, Finland, 8–12 September 2024.
148. Mörbe, W.; Yogeshwar, P.; Tezkan, B.; Kotowski, P.; Thiede, A.; Steuer, A.; Rochlitz, R.; Günther, T.; Brauch, K.; Becken, M. Large-scale 3D inversion of semi-airborne electromagnetic data-topography and induced polarization effects in a graphite exploration scenario. *Geophysics* **2024**, *89*, 1–54. [CrossRef]
149. Smith, R.S.; Koch, R.; Hodges, G.; Lemieux, J. A comparison of airborne electromagnetic data with ground resistivity data over the Midwest deposit in the Athabasca basin. *Near Surf. Geophys.* **2011**, *9*, 319–330. [CrossRef]
150. Pare, P.; Gribenko, A.V.; Cox, L.H.; Čuma, M.; Wilson, G.A.; Zhdanov, M.S.; Legault, J.; Smit, J.; Polome, L. 3D inversion of SPECTREM and ZTEM airborne electromagnetic data from the Pebble Cu–Au–Mo porphyry deposit, Alaska. *Explor. Geophys.* **2012**, *43*, 104–115. [CrossRef]
151. Legault, J.M.; Zhao, S.; Zhao, S. ZTEM airborne AFMAG survey results over low sulphidation epithermal gold-silver vein systems at Gold Springs, south eastern Nevada. *ASEG Ext. Abstr.* **2012**, *2012*, 1–4. [CrossRef]
152. Kowalczyk, P.L.; van Kooten, P.B.M. ZTEM data inversion and interpretation using the UBC-GIF MTinv3D code: A case history at the Silver Queen project, British Columbia. *ASEG Ext. Abstr.* **2012**, *2012*, 1–4. [CrossRef]
153. Orta, M.; Legault, J.M.; Prikhodko, A.; Plastow, G.; Zhao, S.; Ulansky, C. Passive and active helicopter EM survey comparisons over 501 Project Cu- Zn volcanogenic massive sulphide at McFauld's Lake, northern Ontario. *ASEG Ext. Abstr.* **2013**, *2013*, 1–4. [CrossRef]
154. Schijns, H.; Grant, T.; Toole, T. The Hood 10 VHMS deposit of Nunavut, Canada: A case history. *ASEG Ext. Abstr.* **2013**, *2013*, 1–4. [CrossRef]
155. Gaucher, F.; Smith, R. Exploring for copper-gold deposits with time-domain electromagnetics in the Chapais-Chibougamau mining camp: Case history of a challenging variable environment. In Proceedings of the 2016 SEG International Exposition and Annual Meeting, Dallas, TX, USA, 16–21 October 2016; pp. 2228–2232.
156. Legault, J.M.; Latrous, A.; Zhao, S.; Bournas, N.; Plastow, G.C.; Xue, G.G. Helicopter AFMAG (ZTEM) EM and magnetic results over sedimentary exhalative (SEDEX) lead-zinc deposits at Howard's Pass in Selwyn Basin, Yukon. *Explor. Geophys.* **2016**, *47*, 170–178. [CrossRef]
157. Hodges, G.; Chen, T.; van Buren, R. HELITEM detects the Lalor VMS deposit. *Explor. Geophys.* **2016**, *47*, 285–289. [CrossRef]
158. Le, C.V.A.; Harris, B.D.; Pethick, A.M.; Takam Takoumag, E.M.; Howe, B. Semiautomatic and Automatic Cooperative Inversion of Seismic and Magnetotelluric Data. *Surv. Geophys.* **2016**, *37*, 845–896. [CrossRef]
159. Macnae, J. Fitting superparamagnetic and distributed Cole-Cole parameters to airborne electromagnetic data: A case history from Quebec. *Geophysics* **2016**, *81*, B211–B220. [CrossRef]
160. Hübert, J.; Lee, B.M.; Liu, L.; Unsworth, M.J.; Richards, J.P.; Abbassi, B.; Cheng, L.Z.; Oldenburg, D.W.; Legault, J.M.; Rebagliati, M. Three-dimensional imaging of a Ag-Au-rich epithermal system in BritishColumbia, Canada, using airborne z-axis tipper electromagneticand ground-based magnetotelluric data. *Geophysics* **2016**, *81*, B1–B12. [CrossRef]

161. Lee, B.M.; Unsworth, M.J.; Hübert, J.; Richards, J.P.; Legault, J.M. 3D joint inversion of magnetotelluric and airborne tipper data: A case study from the Morrison porphyry Cu–Au–Mo deposit, British Columbia, Canada. *Geophys. Prospect.* **2017**, *66*, 397–421. [CrossRef]
162. Walker, S.E.; Campbell, C.; Legault, J.M.; Izarra, C.; Kwan, K.; Plastow, G.; Whiting, B.; Van Egmond, R. Airborne Geophysics over the Dolly Varden VMS and Low Sulphidation Epithermal Silver Deposits, Northwestern BC, Canada. *ASEG Ext. Abstr.* **2018**, *2018*, 1–5. [CrossRef]
163. Bingham, D.; Legault, J.M. Geophysics of the Patterson Lake South Uranium Deposit, northwestern Saskatchewan. *ASEG Ext. Abstr.* **2018**, *2018*, 1–6. [CrossRef]
164. Lu, X.; Farquharson, C.; Miehé, J.-M.; Harrison, G.; Ledru, P. Computer modeling of electromagnetic data for mineral exploration: Application to uranium exploration in the Athabasca Basin. *Lead. Edge* **2021**, *40*, 139a1–139a11. [CrossRef]
165. Lu, X.; Farquharson, C.G.; Miehé, J.-M.; Harrison. 3D electromagnetic modeling of graphitic faults in the Athabasca Basin using a finite-volume time-domain approach with unstructured grids. *Geophysics* **2021**, *86*, B349–B367. [CrossRef]
166. Smith, R.S.; Lee, T.J. Multiple-order moments of the transient electromagnetic response of a one-dimensional earth with finite conductance—The Gaussian variation applied to a field example. *Explor. Geophys.* **2022**, *53*, 262–274. [CrossRef]
167. Lee, T.J.; Smith, R.S. Multiple-order moments of the transient electromagnetic response of a one-dimensional earth with finite conductance–theory. *Explor. Geophys.* **2021**, *52*, 1–15. [CrossRef]
168. Mir, R.; Fullagar, P.; Darijani, M.; Smith, R.; Scott, S.; Ross, M.; Shamsipour, P.; Chouteau, M.; Ansdell, K.; Gouiza, M. Forward modeling and 3D inversion of electromagnetic data collected over the McArthur River uranium deposit in the Athabasca Basin, Canada. *Geophysics* **2022**, *87*, B129–B143. [CrossRef]
169. La Terra, E.F.; Menezes, P.T.L. Audiomagnetotelluric 3D imaging of the Regis kimberlite pipe, Minas Gerais, Brazil. *J. Appl. Geophys.* **2012**, *77*, 30–38. [CrossRef]
170. McMillan, M.S.; Oldenburg, D.W. Cooperative constrained inversion of multiple electromagnetic data sets. *Geophysics* **2014**, *79*, B173–B185. [CrossRef]
171. Couto Junior, M.A.; Fiandaca, G.; Maurya, P.K.; Christiansen, A.V.; Porsani, J.L.; Auken, E. AEMIP robust inversion using maximum phase angle Cole–Cole model re-parameterisation applied for HTEM survey over Lamego gold mine, Quadrilátero Ferrífero, MG, Brazil. *Explor. Geophys.* **2020**, *51*, 170–183. [CrossRef]
172. Couston, G.; Fitzhardinge, R.; Osborne, G.; Adamson, M. EM success in the Carajas—Geophysical results of the Jaguar Nickel Deposit, Brazil. In Proceedings of the Australasian Exploration Geoscience Conference (AEGC), Brisbane, Australia, 15–17 March 2023.
173. Viezzoli, A.; Kaminski, V. Airborne IP: Examples from the Mount Milligan deposit, Canada, and the Amakinskaya kimberlite pipe, Russia. *Explor. Geophys.* **2016**, *47*, 269–278. [CrossRef]
174. Kaminski, V.; Viezzoli, A. Modeling induced polarization effects in helicopter time-domain electromagnetic data: Field case studies. *Geophysics* **2017**, *82*, B49–B61. [CrossRef]
175. Ghari, H.A.; Voge, M.; Bastani, M.; Pfaffhuber, A.A.; Oskooi, B. Comparing resistivity models from 2D and 1D inversion of frequency domain HEM data over rough terrains: Cases study from Iran and Norway. *Explor. Geophys.* **2020**, *51*, 45–65. [CrossRef]
176. Prikhodko, A.; Bagrianski, A.; Wilson, R.; Belyakov, S.; Esimkhanova, N. Detecting and recovering critical mineral resource systems using broadband total-field airborne natural source audio frequency magnetotellurics measurements. *Geophysics* **2024**, *89*, WB13–WB23. [CrossRef]
177. Thiede, A.; Schiffler, M.; Junge, A.; Becken, M. Multivariate processing of airborne natural source EM data—Application to field data from gobabis (Namibia). *Geophys. J. Int.* **2024**, *238*, 573–589. [CrossRef]
178. Mohanty, W.K.; Mandal, A.; Sharma, S.P.; Gupta, S.; Misra, S. Integrated geological and geophysical studies for delineation of chromite deposits: A case study from Tangarparha, Orissa, India. *Geophysics* **2011**, *76*, B173–B185. [CrossRef]
179. Hoschke, T. Geophysics of the Elang Cu-Au porphyry deposit, Indonesia, and comparison with other Cu-Au porphyry systems. *ASEG Ext. Abstr.* **2012**, *2012*, 1–3. [CrossRef]
180. Legault, J.M.; Izarra, C.; Zhao, S.; Saadawi, E.M. Helicopter AFMAG (ZTEM) Survey Results over the Ad Duwayhi intrusion related gold deposit (IRGD) in the Western Arabian Shield, KSA. *ASEG Ext. Abstr.* **2013**, *2013*, 1–4. [CrossRef]
181. Markandeyulu, A.; Chaturvedi, A.K.; Raju, B.V.S.N.; Parihar, P.S.; Miller, R.; Gooch, G. Application of high resolution airborne geophysical data in geological modelling of Mohar Cauldron Complex, Bundelkhand Massif, central India: Implications for uranium exploration. *Explor. Geophys.* **2014**, *45*, 134–146. [CrossRef]
182. Legault, J.M.; Izarra, C.; Prikhodko, A.; Zhao, S.; Saadawi, E.M. Helicopter EM (ZTEM–VTEM) survey results over the Nuqrah copper–lead–zinc–gold SEDEX massive sulphide deposit in the Western Arabian Shield, Kingdom of Saudi Arabia. *Explor. Geophys.* **2015**, *46*, 36–48. [CrossRef]
183. Xu, S.; Xu, F.; Hu, X.; Zhu, Q.; Zhao, Y.; Liu, S. Electromagnetic characterization of epithermal gold deposits: A case study from the Tuoniuhe gold deposit, Northeast China. *Geophysics* **2020**, *85*, B49–B62. [CrossRef]
184. Chen, J.; Li, Z.; Tian, B.; Chen, X.; Zheng, J.; Zhang, C.; Zhao, W.; Zhang, F. Using the CSAMT method to predict deep mineralisation of copper and molybdenum: A case study of the Zhongxingtun area in Inner Mongolia, China. *Explor. Geophys.* **2020**, *51*, 203–213. [CrossRef]
185. Xu, L.; Yin, Y.; Jin, S.; Wei, W.; Hao, Z.; Li, X.; Liu, H.; Li, H. CSAMT constraints on the metallogenic mechanism of the Zhaishang gold deposit, West Qinling, China. *Explor. Geophys.* **2022**, *53*, 359–371. [CrossRef]

186. Fu, J.; Jia, S.; Wang, E. Combined Magnetic, Transient Electromagnetic, and Magnetotelluric Methods to Detect a BIF-Type Concealed Iron Ore Body: A Case Study in Gongchangling Iron Ore Concentration Area, Southern Liaoning Province, China. *Minerals* **2020**, *10*, 1044. [CrossRef]
187. An, S.; Zhang, Z.; Zhou, K.; Wang, J. Subsurface structures of the Xiaorequanzi deposit, NW China: New insights from gravity, magnetic and electromagnetic data. *Geophys. Prospect.* **2021**, *69*, 434–447.
188. Horo, D.; Pal, S.K.; Singh, S.; Biswas, A. New Insights into the Gold Mineralization in the Babaikundi–Birgaon Axis, North Singhbhum Mobile Belt, Eastern Indian Shield Using Magnetic, Very Low-Frequency Electromagnetic (VLF-EM), and Self-Potential Data. *Minerals* **2023**, *13*, 1289. [CrossRef]
189. Lingaswamy, J.; Thirupathi, K.; Mathur, R.R.; Mukherjee, A. Ground Magnetic and Very Low Frequency Electromagnetic Surveys to Locate Probable Kimberlites in Panna Diamond Belt, Madhya Pradesh, India. *Int. J. Multidiscip. Res.* **2024**, *6*, 1–17.
190. Witherly, K.; Mackee, G. Geophysical Responses over the Cannington Ag-Zn-Pb Deposit-Queensland. *ASEG Ext. Abstr.* **2015**, *2015*, 1–5. [CrossRef]
191. Reid, J.; Farrell, J.; Pears, G.; Jaunzems, A.; Enright, M.; Haederle, M. Integrated Geological and Geophysical Interpretation for the Koodaideri Detrital Iron Deposits, Fortescue Valley, Western Australia. *ASEG Ext. Abstr.* **2016**, *2016*, 1–9. [CrossRef]
192. Jones, H.; Lackie, M.; Smith, M.; Nimalsiri, T. Geophysical Signature of the Southern Gurubang Base Metal Occurrence in South Eastern NSW. *ASEG Ext. Abstr.* **2018**, *2018*, 1–8. [CrossRef]
193. Hoschke, T.; Kim Frankcombe, K. Geophysical response of alteration and mineralisation in the WafiGolpu porphyry system, Papua New Guinea. In Proceedings of the ASEG DISCOVER Symposium (ASEG), Hobart, Australia, 15–18 October 2024.
194. Persson, L.; Lundin, I.A.; Pedersen, L.B.; Claeson, D. Combined magnetic, electromagnetic and resistivity study over a highly conductive formation in Orrivaara, Northern Sweden. *Geophys. Prospect.* **2011**, *59*, 1155–1163. [CrossRef]
195. Le, C.V.A.; Harris, B.D.; Pethick, A.M. Magnetotelluric Inversion, Carbonaceous Phyllites, and an Ore Zone: Kevitsa, Finland. *ASEG Ext. Abstr.* **2016**, *1*, 1–5. [CrossRef]
196. Horrocks, T.; Holden, E.-J.; Wedge, D.; Wijns, C. A nonparametric boundary detection technique applied to 3D inverted surveys of the Kevitsa Ni-Cu-PGE deposit. *Geophysics* **2018**, *83*, IM1–IM13. [CrossRef]
197. Fraser, S.J.; Wilson, G.A.; Cox, L.H.; Čuma, M.; Zhdanov, M.S.; Vallée, M.A. Self-organizing maps for pseudo-lithological classification of 3D airborne electromagnetic, gravity gradiometry and magnetic inversions. *ASEG Ext. Abstr.* **2012**, *2012*, 1–4. [CrossRef]
198. Shah, A.K.; Bedrosian, P.A.; Anderson, E.D.; Kelley, K.D.; Lang, J. Integrated geophysical imaging of a concealed mineral deposit: A case study of the world-class Pebble porphyry deposit in southwestern Alaska. *Geophysics* **2013**, *78*, B317–B328. [CrossRef]
199. Legault, J.M.; Plastow, G.; Zhao, S.; Bournas, N.; Prikhodko, A.; Orta, M. ZTEM and VTEM airborne EM and magnetic results over the Lalor copper-gold volcanogenic massive sulfide deposit region, near Snow Lake, Manitoba. *Interpretation* **2015**, *3*, SL83–SL94. [CrossRef]
200. Yang, D.; Oldenburg, D.W. 3D inversion of total magnetic intensity data for time-domain EM at the Lalor massive sulfide deposit. *Explor. Geophys.* **2017**, *48*, 110–123. [CrossRef]
201. Yang, D.; Fournier, D.; Kang, S.; Oldenburg, D.W. Deep mineral exploration using multi-scale electromagnetic geophysics: The Lalor massive sulfide deposit case study. *Can. J. Earth Sci.* **2018**, *56*, 544–555. [CrossRef]
202. Hayward, N.; Corriveau, L.; Craven, J.A.; Enkin, R.J. Geophysical Signature of the NICO Au-Co-Bi-Cu Deposit and Its Iron Oxide-Alkali Alteration System, Northwest Territories, Canada. *Econ. Geol.* **2016**, *111*, 2087–2109. [CrossRef]
203. Peacock, J.R.; Denton, K.M.; Ponce, D.A. Magnetotelluric Imaging of a Carbonatite Terrane in the Southeast Mojave Desert, California and Nevada. *ASEG Ext. Abstr.* **2016**, *2016*, 1–5. [CrossRef]
204. Kaminski, V.; Di Massa, D.; Viezzoli, A. Joint inversions of two VTEM surveys using quasi-3D TDEM and 3D magnetic inversion algorithms. *Explor. Geophys.* **2016**, *47*, 260–268. [CrossRef]
205. Devriese, S.G.R.; Davis, K.; Oldenburg, D.W. Inversion of airborne geophysics over the DO-27/DO-18 kimberlites—Part 1: Potential fields. *Interpretation* **2017**, *5*, T299–T311. [CrossRef]
206. Fournier, D.; Kang, S.; McMillan, M.S.; Oldenburg, D.W. Inversion of airborne geophysics over the DO-27/DO-18 kimberlites—Part 2: Electromagnetics. *Interpretation* **2017**, *5*, T313–T325. [CrossRef]
207. Kang, S.; Fournier, D.; Oldenburg, D.W. Inversion of airborne geophysics over the DO-27/DO-18 kimberlites—Part 3: Induced polarization. *Interpretation* **2017**, *5*, T327–T340. [CrossRef]
208. Mir, R.; Perrouty, S.; Astic, T.; Bérubé, C.L.; Smith, R.S. Structural complexity inferred from anisotropic resistivity: Example from airborne EM and compilation of historical resistivity/induced polarization data from the gold-rich Canadian Malartic district, Québec, Canada. *Geophysics* **2019**, *84*, B153–B167. [CrossRef]
209. Kuhn, S.; Cracknell, M.J.; Reading, A.M.; Sykora, S. Identification of intrusive lithologies in volcanogenic terrains in British Columbia by machine learning using random forests: The value of using a soft classifier. *Geophysics* **2020**, *85*, B249–B258. [CrossRef]
210. Darijani, M.; Farquharson, C.G. Inversion of magnetic and frequency-domain electromagnetic data for investigating lithologies associated with gold mineralization in the Canadian Malartic area, Québec, Canada. *Can. J. Earth Sci.* **2021**, *58*, 413–432. [CrossRef]
211. Finn, C.A.; Zientek, M.; Bloss, B.R.; Parks, H.; Modroo, J. Electromagnetic and magnetic imaging of the Stillwater Complex, Montana, USA. *Explor. Geophys.* **2023**, *54*, 553–570. [CrossRef]

212. Legault, J.M.; Kwan, K.; Greig, J.; Webster, E.; Hanki, M. Targeting epithermal Au-Ag using helicopter TDEM, magnetic, and radiometric data at Lawyers Project, North-Central BC, Canada. In Proceedings of the 8th International Airborne Electromagnetics Workshop (AEM2023), Fitzroy Island, Australia, 4–7 September 2023.
213. Legault, J.M.; Kwan, K.; Ebert, S. ZTEM Airborne Natural Field EM-Magnetics and Mineral Targeting Results over the Berg Porphyry Copper Project, near Houston, British Columbia. In Proceedings of the 8th International Airborne Electromagnetics Workshop (AEM2023), Fitzroy Island, Australia, 4–7 September 2023.
214. Legault, J.M.; Niemi, J.; Brett, J.; Zhao, S.; Han, Z.; Plastow, G. Passive airborne EM and Ground IP\Resistivity results over the Romero intermediate sulphidation epithermal gold deposits, Dominican Republic. *Explor. Geophys.* **2016**, *47*, 191–200. [CrossRef]
215. Kwan, K.; Plastow, G.; Prikhodko, A.; Kapetas, J.; Legault, J.M.; Druecker, M. VTEM airborne EM, aeromagnetic and gamma-ray spectrometric data over the Cerro Quema high sulphidation epithermal gold deposits, Panama. *ASEG Ext. Abstr.* **2015**, *2015*, 1–4. [CrossRef]
216. Legault, J.M.; Wijns, C.; Izarra, C.; Plastow, G. The Balboa ZTEM Cu-Mo-Au porphyry discovery at Cobre Panama. *ASEG Ext. Abstr.* **2016**, *2016*, 1–6.

MDPI AG
Grosspeteranlage 5
4052 Basel
Switzerland
Tel.: +41 61 683 77 34

Minerals Editorial Office
E-mail: minerals@mdpi.com
www.mdpi.com/journal/minerals